W0090779

Industrial Aspects of
Biochemistry and Genetics

NATO ASI Series

Advanced Science Institutes Series

A series presenting the results of activities sponsored by the NATO Science Committee, which aims at the dissemination of advanced scientific and technological knowledge, with a view to strengthening links between scientific communities.

The series is published by an international board of publishers in conjunction with the NATO Scientific Affairs Division

A	**Life Sciences**	Plenum Publishing Corporation
B	**Physics**	New York and London
C	**Mathematical and Physical Sciences**	D. Reidel Publishing Company Dordrecht, Boston, and Lancaster
D	**Behavioral and Social Sciences**	Martinus Nijhoff Publishers
E	**Engineering and Materials Sciences**	The Hague, Boston, and Lancaster
F	**Computer and Systems Sciences**	Springer-Verlag
G	**Ecological Sciences**	Berlin, Heidelberg, New York, and Tokyo

Recent Volumes in this Series

Series A: Life Sciences

Industrial Aspects of Biochemistry and Genetics

Edited by

N. Gürdal Alaeddinoğlu

Middle East Technical University
Ankara, Turkey

Arnold L. Demain

Massachusetts Institute of Technology
Cambridge, Massachusetts

and

Giancarlo Lancini

Lepetit Research Laboratories
Milan, Italy

Plenum Press
New York and London
Published in cooperation with NATO Scientific Affairs Division

Proceedings of a NATO Advanced Study Institute on
Industrial Aspects of Biochemistry and Genetics,
held September 4–14, 1983,
in Cesme, Turkey

ISBN 978-1-4684-1229-1   ISBN 978-1-4684-1227-7 (eBook)
DOI 10.1007/978-1-4684-1227-7

Library of Congress Cataloging in Publication Data

NATO Advanced Study Institute on Industrial Aspects of Biochemistry and Genetics
(1983: Çesme, Turkey)
Industrial aspects of biochemistry and genetics.

(NATO ASI series. Series A, Life sciences; vol. 87)
"Proceedings of a NATO Advanced Study Institute on Industrial Aspects of Bio-
chemistry and Genetics, held September 4–14, 1983, in Cesme, Turkey"—T.p. verso.
"Published in cooperation with NATO Scientific Affairs Division."
Bibliography: p.
Includes index.
1. Industrial microbiology—Congresses. 2. Biochemical engineering—Congress-
es. 3. Microbial genetics—Congresses. I. Alaeddinoğlu, N. Gürdal. II. North Atlantic
Treaty Organization. Scientific Affairs Division. III. Title. IV. Series: NATO ASI series.
Series A, Life sciences; v. 87.
QR53.N37   1983   660.6   85-3690
ISBN 978-1-4684-1229-1

©1985 Plenum Press, New York
Softcover reprint of the hardcover 1st edition 1985

A Division of Plenum Publishing Corporation
233 Spring Street, New York, N.Y. 10013

All rights reserved

No part of this book may be reproduced, stored in a retrieval system, or transmitted,
in any form or by any means, electronic, mechanical, photocopying, microfilming,
recording, or otherwise, without written permission from the Publisher

PREFACE

This book includes the proceedings of a NATO Advanced Study
Institute held at Cesme (Turkey), from September 4 to 14, 1983.

Recent advances in molecular biology have generated a wave of
excitement about the prospective application of novel microbiological techniques in a wide range of industrial roles. The new
methodology, brought about by recombinant DNA technology, has given
the biologist direct access to the genome which in turn, envisages
the adaption of certain microorganisms to industrial production,
much higher yields of such products as antibiotics and enzymes by
biotechnological industry, and the industrial production of substances which occur in nature in very low concentrations and cannot
be economically recovered.

In the chapters that follow, the accomplishments and the
promise of genetic engineering and industrial microbiology are set
forth by its practitioners.

We greatly acknowledge the help of those in preparation of the
individual manuscripts and our thanks to all participants of the
meeting for their particular efforts.

Nafi G Alaeddinoğlu
Arnold L Demain
Giancarlo C Lancini

CONTENTS

PLASMID INCOMPATIBILITY AND REPLICATION

Aslihan Tolun

Biology Department
Boğaziçi University
Bebek, Istanbul

Plasmids are covalently closed circular DNA molecules that replicate autonomously in bacteria. Because of their small size and relative simplicity, their replication was studied extensively as a model for the replication of the complex cell chromosome. Recently, plasmids have gained more attention because they are the most commonly used cloning vehicles in genetic engineering experiments. The earliest gene cloning experiments were done solely for scientific interest, but soon the contribution of gene cloning techniques to industry was realized and are now being used extensively in biotechnology.

Some plasmids, namely the R factors, confer antibiotic resistance to the host bacterium. These plasmids are important in the epidemiology of antibiotic resistant pathogens of both man and animals. A special property of some of the R factors is self-transmissibility from the host to another bacterium. This property results in the spread of antibiotic resistance among bacterial pathogens (1).

Studying the relatedness of plasmids in pathogenic bacteria can reveal the epidemiology of drug resistant clinical isolates. Relatedness of plasmids can now be studied by DNA homology, but the faster and simpler traditional method has been by examining incompatibility properties. Two plasmids are said to be incompatible if they cannot be stably inherited in a cell line. If a bacterium is transformed with two such plasmids simultaneously, after some cell divisions, each bacterium in the culture will harbor only one or the other plasmid, but not both. This is the result of competition for similar factors necessary for replication and maintenance. Plasmids can be classified by placing those which are incompatible in the same group. Incompatible plasmids indeed exhibit similar replication properties such as copy number and have DNA sequences in common. However, since a plasmid

1

may be incompatible with members of more than one group, classification of plasmids by incompatibility properties is an oversimplification. Nevertheless, this classification gives a broad idea about plasmid replication and is important in studies of the spread of clinically important plasmids. Incompatibility properties of mainly the E.coli plasmids have been studied in detail. These plasmids constitute about 25 main incompatibility groups, and members of each group exhibit properties similar with each other, but quite different than those of other groups (1). Studies of the mechanisms of incompatibility have generated much information about plasmid maintenance which has been useful for the construction of cloning vehicles. The general features of maintenance which will be discussed are stable inheritance/partitioning and autonomous replication with a defined copy number.

At least for low copy number plasmids, a partitioning mechanism exists ensuring that each new cell inherits at least one copy of the plasmid. For high copy number plasmids segregation to daughter cells may not need any active partitioning mechanism. Because there are a large number of plasmid copies in a cell, the probability for each daughter cell to inherit at least one copy is high. However, this random segregation model is not feasible for low copy number plasmids. For example F plasmid which exists as 1-2 copies per cell, is very stable in a cell line. That degree of stability would not be expected if segregationn were random. The alternative model to random segregation is regulated segregation, or equipartition which presumes that an active mechanism ensures that each daughter cell inherits at least one copy of the plasmid. Since F plasmid partitioning is so finely controlled, an obvious assumption is that its partitioning is controlled by a mechanism similar to that of the cell chromosome. The present data is not sufficient to clarify this point. The equipartition model presumes the existance of partitioning genes; the incD region of F plasmid may provide this function. A region of DNA is considered to be an incompatibility region (inc) if it prevents the maintenance of the parent plasmid in the same cell when it is joined to an unrelated plasmid. Mini F is a 9 kilobase (kb) EcoRI fragment of the 94.5 kb F plasmid which is capable of autonomous replication; it also exhibits the same replication properties of F plasmid (2,3). This method of dissection of plasmids to smaller autonomously replicating fragments has been very useful for studying DNA replication. IncD region is not essential for plasmid replication, but its deletion results in an unstable plasmid (4). Among the three inc regions of F, incD expresses the weakest incompatibility against F plasmids (5). IncD is in the region common to other plasmids in the FI incompatibility group, as has been shown by DNA heteroduplex mapping using electron microscopy (6). It is also the only inc region that expresses incompatibility against those plasmids (7).

To find out the nature of the partitioning gene, Ogura and Hiraga mapped three separate genes in the incD region (8). SopA (sop:stabi-

2

lity of plasmid) and sopB act in trans, but sopC functions only in cis, suggesting that it is the specific site necessary for actual partitioning. The data also suggest that the proteins coded by sopA and sopB may act on sopC. Hayakawa and Matsubara have found that the sopB protein binds around sopC; two presumably host-coded proteins bind sopC concomitantly. The sopC gene product in conjunction with these host proteins may therefore bind to the partitioning site on cell membrane. Irrespective of whether the proposed model proves to be correct or not, the present data on incD offers strong evidence that partitioning genes exists, and one way that plasmid incompatibility arises is by competition for stability.

The only other inc regions yet described, incB and incC, play a role in replication and copy number control (9). It contributes greatly to incompatibility by keeping the total number of copies fixed. If in the same cell there are two similar plasmids, A and B, each having a copy number of two, then after plasmid replication and cell division each daughter cell will inherit any two copies. Since it cannot distinguish between the two plasmids, two possibilities with equal probability emerge: the two daughter cells harbor two copies of one plasmid or the two daughter cells harbor one of each plasmid. For the latter cells, there is again only 50% chance that they will give rise to cells with mixed plasmids. Therefore, eventually each cell in the culture will harbor only one or the other plasmid. For higher copy number plasmids the picture is similar, but of course more cell doublings are necessary to reach the pure plasmid state.

Two models have been proposed to explain how the copy number can be kept at a particular level. The positive control model proposed by Jacob et.al. (10) suggests that the number of copies of a plasmid is determined by the limited amount of a factor (or factors) necessary for replication. Alternatively, the negative control model proposed by Pritchard et.al. (13) assumes that there is a diffusable inhibitor of replication that acts in trans and is coded by the plasmid. The concentration of the inhibitor is proportional to the number of plasmids in the cell, and is diluted out as the cell mass increases before cell division. The currently available data suggest that a number of plasmids such as ColE1 employ the latter model. A 555 nucleotide long primer RNA for initiation of ColE1 DNA replication is processed by RNAseH from the primer precursor transcribed from the region to the right of the origin of DNA replication (ori). A 108 nucleotide long transcript synthesized from an overlapping segment of the complementary strand is partially homologous to primer RNA. This RNA I transcript interacts with primer RNA to inhibit the initiation of DNA replication (12). The rop gene located to the left of ori also regulates DNA replication by inhibiting the initiation of transcription of primer RNA (13). This gene codes for a 63 amino acid protein which is not essential for replication. Of the three genes, only RNAI influences incompatibility. Lacetena and Cesareni (14) and Tomizawa and Itoh (15) showed independently that inhibition is due to base pairing of RNAI with the primer RNA, thus inhibiting base pairing of the primer with the DNA. Mutations which reduce the homology (hybridization) between RNAI and the primer RNA increase the availability of primer RNA for the initiation of DNA replication. This results in a decrease in the expression of incompatibility and an increase in copy number.

It is not known why two negative control systems, RNAI and the rop gene, have evolved in ColE1 plasmid. It is interesting that of these two genes only the RNAI plays a role in the expression of incompatibility. ColE1 replication is nevertheless a simple system that does not even require any proteins coded by the plasmid, and it exhibits probably the simplest incompatibility mechanism.

For larger plasmids, the picture seems much more complicated, and not yet clear. For example, for plasmids RI (a large plasmid with low copy number), two regions (cop) control the copy number but only copA expresses incompatibility. The product of this gene was shown by Stougaard et al (16) to be a 90 nucleotide long unstable RNA, with a half life of less than a few minutes. The nucleotide sequence analysis revealed that this RNA is like RNAI of ColE1 in that it is untranslatable, but has potential for high degree of secondary structure. Light and Molin (17) showed that the target of this RNA lies between the promoter and a gene which is required for replication (repA). Thus copA RNA prevents expression of repA gene, either by interfering with translation or with transcription. The mechanism of action of this RNA is therefore very different than that of ColE1 RNAI, in spite of the similarity of function.

Our present knowledge of F plasmid replication is limited. As previously mentioned, copy number of this plasmid is strictly controlled. Between the regions incB and incC is the gene for the 29K protein essential for replication (18). Kline (16) has shown that disruption of the incC region by transposon insertion results in high copy number. Five 22 base pair (bp) direct repeats are found in the nucleotide sequence of this region (18, 19). The presence of four similar 19-22 bp repeats in incB was revealed by Murotsu et.al. (18). When cloned into another plasmid, a 58 bp fragment of incC containing two of the repeats expresses incompatibility against F plasmid. Two such fragments cloned in tandem express much stronger incompatibility (19). What these repeats do at the molecular level is still not known, but similar repeats are observed in the replication regions of other plasmids as well as bacteriophage lambda (figure 1). These are the repeats in the region essential for replication of pR6K, a 38 kb plasmid that specifies resistance to antibiotics ampicillin and streptomycin. The repeat indicated by asterisk is in the promoter region for the protein that initiates DNA replication in vitro (20). The sequence analysis has revealed the presence of eight similar repeats in pRK2 (21). RK2 is a 56 kb plasmid with copy number of seven and is able to replicate in a variety of hosts. The repeats have a common core sequence, a hexanucleotide, except in the case of pR6K where G has been replaced by the other purine. The core sequence is also found in the lambda phage replication region (22). It is also found once in the replication region of E.coli chromosome (23,24). Some plasmids such as RI, however, do not contain repeats in replication regions. With the available data, we can only speculate on what the repeats do. They may serve as recognition sites for either the membrane or for a protein necessary for replication. Alternatively, they may code for a small repressor RNA as in the case of ColE1 and RI plasmids. A more surprising answer is of course possible.

I would like to draw your attention to F plasmid again. As I mentioned earlier the incD region is common to other plasmids in FI group and is the only region that expresses FI incompatibility. Why do the other inc genes not express incompatibility? DNA heteroduplex mapping using electron microscopy showed that the region containing these genes does not have good homology with DNA of other FI group

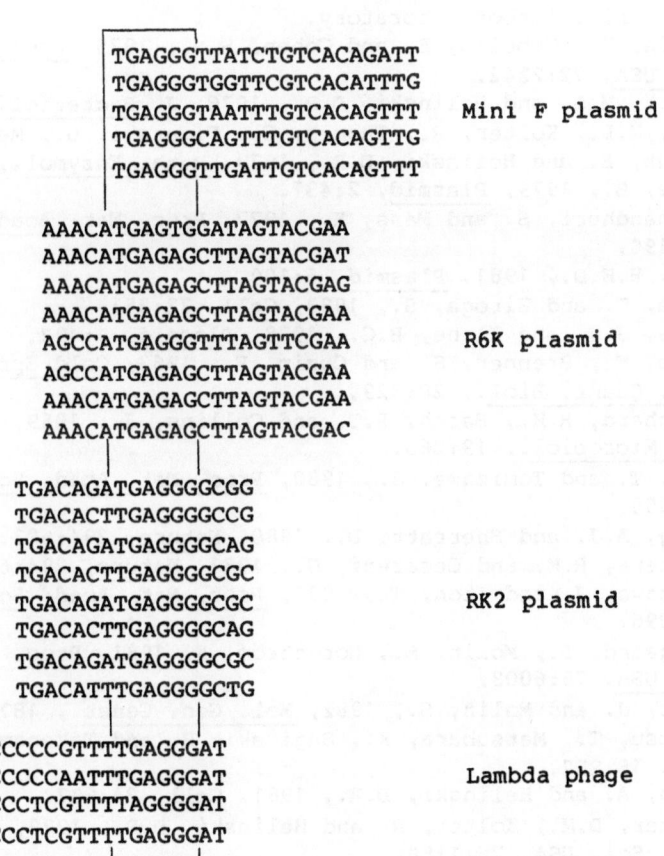

Figure 1. Families of direct repeat units aligned by a box to show the presence of a 6 bp homologous sequence.

plasmids (6). It was interesting to find out whether repeats similar to those in incB and incC were present in other plasmids in the FI group. Southern blot hybridization have shown that all of the four plasmids tested have sequences rather similar, but not identical with the incC repeats (A. Tolun, unpublished observations). Therefore, the overall replication models of the FI plasmids are most probably similar, even though the incB and the incC regions have not been conserved as well as the incD region during the course of evolution. Studies of DNA homology have thus confirmed the classification of plasmids by incompatibility.

References

1. Jacob, A.E., Shapiro, J.A., Yamamoto, L., Smith, D.I., Cohen, S.N. and Berg, D.: Plasmids Studied in Escherichia coli and Other Enteric Bacteria, p 607 in "DNA Insertion Elements, Plasmids, and Episomes" (1977) Eds: A.I. Bukhari, J.A. Shapiro, S.L. Adhya. Cold Spring Harbor Laboratory.
2. Timmis, K., Cabello, F. and Cohen, S.N., 1975, Proc. Nat. Acad. Sci. USA, 72:2242.
3. Lovett, M.A. and Helinski, D.R., 1976, J. Bacteriol., 127:982.
4. Kahn, M.L., Kolter, R., Thomas, C., Figurski, D., Meyer, R., Remaut, E. and Helinski, D.R., 1979, Meth. Enzymol., 68:368.
5. Kline, B., 1979, Plasmid, 2:437.
6. Palchandhuri, S. and Maas, K., 1977, Proc. Nat. Acad. Sci. USA, 74:1190.
7. Lane, H.E.D., 1981, Plasmid, 5:100.
8. Ogura, T. and Hiroga, S., 1983, Cell, 32:351.
9. Manis, J.J. and Kline, B.C., 1978, Plasmid, 1:492.
10. Jacob, F., Brenner, S. and Cuzin, F., 1963, Cold Spring Harbor Symp. Quant. Biol., 28:329.
11. Pritchard, R.H., Barth, P.T. and Collins, J., 1969, Symp. Soc. Gen. Microbiol., 19:263.
12. Itoh, T. and Tomizawa, J., 1980, Proc. Nat. Acad. Sci. USA, 77:2450.
13. Twigg, A.J. and Sherratt, D., 1980, Nature, 294:623.
14. Lacatena, R.M. and Cesareni, G., 1981, Nature, 294:623.
15. Tomizawa, J. and Itoh, T., 1981, Proc. Nat. Acad. Sci. USA, 78:6096.
16. Stougaard, P., Molin, S., Nordström, K, 1981, Proc. Nat. Acad. Sci. USA, 78:6008.
17. Light, J. and Molin, S., 1982, Mol. Gen. Genet., 187:486.
18. Murotsu, T., Matsubara, K., Sugisaki, H. and Takanami, M., 1981, Gene, 15:257.
19. Tolun, A. and Helinski, D.R., 1981, Cell, 24:687.
20. Stalker, D.M., Kolter, R. and Helinski, D.R., 1979, Proc. Nat. Acad. Sci. USA, 76:1150.
21. Stalker, D.M., Thomas, C.M. and Helinski, D.R., 1981, Mol. Gen. Genet., 181:8.

22. Grosschedl, R. and Hobom, G., 1979, Nature, 277:621.
23. Horota, Y., Yasuda, S., Yamada, M., Nishimura, A., Sugimoto, K., Sugisaki, H., Oka, A. and Takanami, M., 1978, Cold Spring Harbor Symp. Quant. Biol., 78:129.
24. Messer, W., Meijer, M., Bergmans, H.E.N., Hansen, F.G., Meyenburg von, K., Beck, E. and Schaller, H., 1978, Cold Spring Harbor Symp. Quant. Biol., 78:139.

A TRANSPOSABLE ELEMENT FROM <u>HALOBACTERIUM</u> <u>HALOBIUM</u> WHICH

INACTIVATES THE BACTERIORHODOPSIN GENE

Mehmet Şimşek

Middle East Technical University
Department of Biological Sciences
Ankara, Turkey

INTRODUCTION

   <u>Halobacterium</u> <u>halobium</u> belongs to a class of organisms termed
"Archaebacteria" by Woose and his coworkers (1). This bacterium
displays both prokaryotic and eukaryotic features and is proposed
to constitute a major line of descent besides prokaryotes and
eukaryotes. One interesting feature of <u>H.halobium</u> is its ability
to utilize light energy without involvement of photosynthesis. It
is known that light transduction is achieved by means of a membrane
protein (bacteriorhodopsin) which catalyzes the light dependent
translocation of protons out of cells and hence produces a trans-
membrane electrochemical gradient. This gradient is not wasted but
is used by cells to synthesize ATP and to drive other transport
processes (2).

   There are, however, spontaneous mutants (Pum^-) of <u>H.halobium</u>
which cannot utilize light energy because they lack the protein
(bacteriorhodopsin, <u>BR</u>) in their membrane. We have recently cloned
and characterized the <u>BR</u> gene from such two mutants and shown (3)
that absence of bacteriorhodopsin is due to inactivation of the
gene by insertion of a 1.1 Kbp sequence near the NH_2-terminal end.
Further characterization of the 1.1 Kbp insertion sequence revealed

---

Abbreviations. <u>BR</u>, bacteriorhodopsin ; $\underline{Pum}^-$, spontaneous mutant
of <u>H.halobium</u> which can not synthesize BR ; IS, insertion sequence
ISH, insertion sequence from <u>H.halobium</u> ; Kbp, kilo base pairs ;
Tn , Transposable element ; RF, replicative form ; ORF, open
reading frame.

that it shares the features common to transportable IS elements (4).
Thus, it is termed ISH₁ and represents the first transposable
element to be characterized from halobacteria. The complete sequence
of ISH1 has been determined (3). It is 1,118 nucleotides long and
contains an interrupted 8 bp inverted repeat at its ends. Like other
IS elements, it duplicates an 8 bp target sequence upon insertion.
Analysis of various strains of H.halobium showed that the copy number
of ISH1 ranges from one to five or more in different strains.

CLONING OF TWO MUTANT BACTERIORHODOPSIN GENES THAT CONTAIN
A 1.1 kbp INSERT (ISH1) WITHIN THE GENE

Wild type BR gene was previously cloned (5) from H. halobium strain
S9 into the Pst 1 site of pBR₃₂₂. In further attempts to generate
a Bam H1 clone bank from the same halobacterium strain, a clone
designated pMSb1, was obtained (3). In this clone, the BR gene was
found to be on a 7.5 kbp Bam H1 fragment (Fig.1, lane 2) which is
approximately 1.1 kbp longer than the BamH1 fragment of the strain
S9 that contains the intact BR gene (Fig.1C). Thus, it was likely
that pMSb 1 represented the cloning of the BR gene from a spontaneous
Pum mutant in the S9 culture. In fact, such a spontaneous mutant,
SD17, that can not produce bacteriorhodopsin was isolated later
on and upon cloning of the BR gene from this mutant, the BR

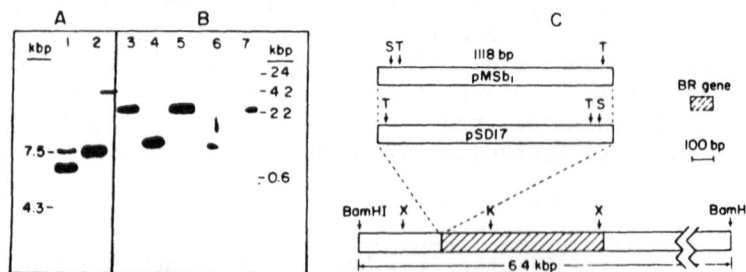

Figure 1. Identification of BR gene in the restriction digests of
genomic and cloned DNAs from various H.halobium strains. (A),
Southern hybridization of BR cDNA probe to Bam H1 digests of genomic
DNA from S9 (lane 1) and pMSb1 (lane 2). (B) Southern hybridization
of BR cDNA probe to Xmalll digests of DNA's from pMSb1 (lane 3), pR1
(lane 4), pSD17 (lane 5), R1 (lane 6) and SD17 (lane 7). Numbers on
the sides indicate DNA size markers. (C), Partial restriction map of
Bam H1 fragments containing the BR gene in different clones. The
wild type BR gene (hatched area) is on a 6.4 kbp BamH1 fragment.
T, Tth1 ; S, Smal ; X, Xmalll ; and K, Kpnl.

10

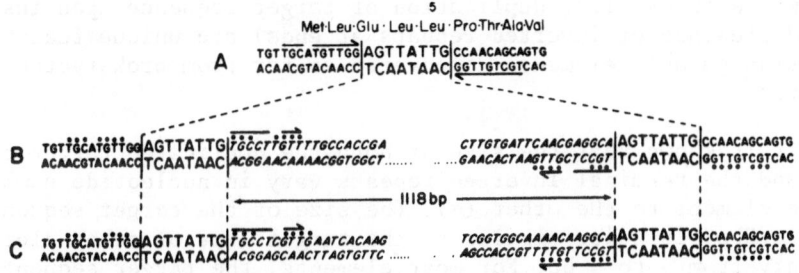

Figure 2. Junction sequences at the insertion site of ISH1 within the BR gene. (A), Partial DNA sequence for the NH$_2$-terminus of the BR gene. (B and C), BR gene-ISH1 junction sequences in pMSb1 and pSD17 DNAs, respectively. Nucleotides written in larger letters correspond to the 8 bp target sequence at the insertion site (A) which is duplicated and flanks the inserted ISH1 as direct repeats (B and C). Inverted repeat sequences in the BR gene and at the ends of ISH1 are indicated by half arrows. Dots indicate homology in DNA sequence between the BR gene and ends of ISH1.

gene again was found on a 7.5 kbp Bam H1 fragment (Fig.1C). Comparison of pMSb1 and pSD17 by restriction mapping showed that they both contain a 1.1 kbp insert within the BR gene (Fig.1B,1C). Further mapping using Kpn1 (data not shown) localized the 1.1 kbp insertion between the Xma 111 and Kpn1 sites (Fig.1C).

CHARACTERIZATION OF THE 1.1 Kbp INSERT (ISH1) BY SEQUENCE ANALYSIS

In order to determine more precisely the location of ISH1 element in the Xma 111-KpnI region, sequence analysis of both pMSb1 and pSD17 were carried out around the junction areas using the method of Maxam and Gilbert (6). For the pMSb1, the SmaI site was used for determining the sequence at the left junction and the Kpn1 site for the right junction (Fig.1C). Similarly, for the pSD17, the junctions were determined by sequencing at the Tth1 and SmaI sites (Fig.1C, T and S at each end of ISH1 in pSD17).

Fig.2 shows the site of insertion of ISH1 and the junctional sequences between the BR gene and the ISH1 elements in pMSb1 (Fig.2B) and in pSD17 (Fig.2C). It is interesting to see that the ISH1 element (1.1kbp) is inserted into an identical position of the BR gene for both pMSb1 and pSD17, but the insertion is in reverse orientation. Furthermore, insertion of ISH1 leads to the duplication of an 8-bp target sequence (A-G-T-T-A-T-T-G.) and this duplicated sequence is found as a direct repeat flanking each end of ISH1. The ends of ISH1 itself contain an interrupted 8 bp inverted repeat(T-G-C-C-T-.-G-T-T-).

These two features (i.e. duplication of target sequence upon insertion and presence of inverted repeats at ends) are unique features of all transposable elements discovered so far from prokaryotes and eukaryotes (4).

It is known that the size of both the duplicated target sequence and the terminal inverted repeats vary in nucleotide number from one element to the other (4). The size of the target sequence for ISH1 (8bp) is quite similar to the target size of other elements which vary from 5 to 9 bp. For most elements, the target sequences have no significant homology and the insertions appears to be at random, whereas the elements such as IS4, Tn554, Tn7, Tn10 and the copia-like element 297 of Drosophila are quite sequence specific for insertions (7). In a few cases (8) (e.g. Tn 10 and copia-like Drosophila elements) the insertions occur within a symmetrical target sequence. In the case of ISH1, duplicated target sequence (A-G-T-T-A-T-T-G) is not symmetrical. However, it is interesting that this target sequence is flanked by an almost perfect 9-bp inverted repeat in the BR gene (Fig.2A, opposite arrows). Furthermore, the nucleotides within this inverted repeat show significant homology to the ends of ISH1. A similar situation exists also in the case of Tn7 which inserts at a unique site on the E.coli chromosome and Col E1 DNA (9). Whether ISH1 insert into H.halobium DNA in a site-specific manner and whether the sequence feature around the target site contribute to the specifity of insertion is not known at present and awaits cloning and junction sequence analysis of other halobacterial DNA fragments with ISH1 insertions. Recently, DasSarma et al. (10) isolated four pum H.halobium mutants which can not synthesize bacteriorhodopsins and the deficiency in these mutants is shown to be due to insertion of the ISH1 element within the BR gene at exactly the same target site as in pMSb1 and pSD17 in this study. Thus, at least for the BR gene, insertions of ISH1 appear to be quite site specific.

Complete Sequence of ISH1

In order to get more information about the ISH1 element and compare it with the established sequences of known bacterial IS-elements, complete sequencing of the ISH1 in pMSb1 was under-taken using the M13 cloning and sequencing system (11). For this, a 1.8 kbp BamHl-Kpnl fragment of pMSb1 (Fig.1C) was isolated and cleaved seperately with Sau 3A, Taql and HpaII. The resulting fragments were inserted into either the BamHl or the AccI site of M13 mp7 RF DNA for cloning. The single stranded recombinant M13 DNAs were isolated from the colourless phage plaques and used as template for the sequence analysis using the dideoxy chain termination method of Sanger et al. (11).

Fig.3 shows the complete sequence of ISH1. Comparison of this sequence with the eubacterial IS elements (IS1, IS2, IS5) revealed no significant homology(12-14) ISH1 was also compared with two

```
                                                                          60
TGCCTTGTTT TGCCACCGAT TGAGGGAAGT TTCAGACTCT CTCCCGGGAA GATTCCGTCA
                                                                         120
AGCTAACCAG GAATTGGACG CCGTCTGGCG ATATGGCATC GCTCAGACGG CTTGCTTGGA
                                                                         180
TGTGTCGAAA CCTTGCCAAA CAGCACGTTG ACGATCCGGA CGTACCCGCC GCGCCGTCCG
                                                                         240
GCGCGGGCGG GTACGCCGAG TGGGTGCAGA TCGCCGTTGAT TCTGTACCGT GTCGAACTGG
                                                                         300
AAAAGAGCCT CCGTGAATCC GAGGACTACC TCAACGAGAT GCCCGGTGTT CTTGCCGTGT
                                                                         360
TTGGACTTGA CGAAGCACCA CACTACAGCT CGTTCTGCCG GTGGGAAAAC GAGTATCGAA
                                                                         420
TGCCGTGAGCT CCGCCGCCTG CTCCCGCGCTT CGGCGGAGCA GGCGGGCTGG AGTGGCGAAG
                                                                         480
CCGCGATTGA CGCGAGCGGC TTCCAGCGCG ATCAAACCAG CTACCACTAC CGCGACCGCG
                                                                         540
CGAATTACTC GTTCCAGTCG ATGAAGACGA CGATCTTGAT CGACGTGAAC TCGCTAGCGA
                                                                         600
TCAAGGACGT TCACTACACG ACGCAGAAGC CTGGGACGGC CACATTGGGA TGCAGGTCTT
                                                                         660
CCGCCGGAAA' CGCGGAAGAC CTGCGGGTGC TGTCTGCTGA CGCGAACTAC TCGTGGAGCG
                                                                         720
ACCTCCGTGA GGAGTGTCGC TCCGAATCAA CGCGACCGTT GATCAAGCAC AGGGAGCAAA
                                                                         780
CACCGTTGCA GAAGGCTCAC CACGCCCGGA TGAACGAGGA CTACAACCAA CGCTGGATGA
                                                                         840
GTGAAACCGG CTTCTCGCAG TTGAAGGAAG ACGACGGCGA GAAGCTGCGC TCCGGAGCTG
                                                                         900
GCAGGCCAGT TCCGGAGCTG ACTCGGAAGT GCATCATCCA TAACCTGACG CAGGCGGCGA
                                                                         960
GTTAAGGGCT CGCCGCCTGC TCGCTTTCTC CGTACGTATC CGGAGAGGCA TCGCCGTCGT
                                                                        1020
CATCGGAACA ACGAAGCAAG ATACCATAGT TGTGACCCTT CAGCAACCGC CGTGAGTGAC
                                                                        1080
AGCTACTGCA TCTTCTGAGG TCAAGAACCC GTCTCTGACG CTGTGAAACT GCGAATAGTC

TTCCCTACCC CGACGCTGTC TTGTGATTCA ACGAGGCA
```

Figure 3. Complete nucleotide sequence of 1SH1 in pMSb1 DNA. The sequence is of the strand which contains the long ORF (nucleotides 93–902)* Opposite half arrows indicate internal inverted repeat sequences which are eight nucleotides or longer in length.

recently discovered(15, 16)halobacterial IS elements (ISH2 and ISH50). Again, there was no significant homology. However, one common feature of these IS elements is the presence of a long open reading frame (ORF) which is believed to encode a protein that is required for the transposition of that element. ISH1 has such a long ORF which starts with the ATG codon at position 92 and ends with the TAA codon at position 902 (Fig.3). This ORF (810 nucleotides) can code for a polypeptide of 270 aminoacids. If GUG is used as an initiator codon in halobacteria, the sequence complementary to that shown in Fig.3 also has a relatively long ORF (nucleotides 467 to 66) which is contained almost totally within the longer ORF in the other strand. This situation is analogous to that observed for several E.coli IS elements in which both strands of an element have ORFs. A similar situation has also been reported recently for the halobacterial ISH50 element (16) which has a long (819 nucleotide) ORF in one strand and a shorter (366 nucleotide) ORF in the other strand.

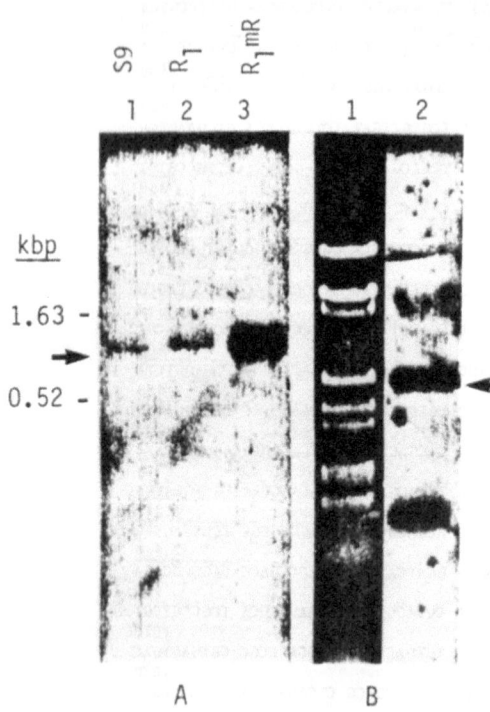

S9 R₁ R₁mR

Figure 4. Detection of transcripts corresponding to ISH1 in H.halobium. (A), RNA gel transfer hybridization of RNAs isolated from strains S9(lane 1), RI(lane 2), and R1mR(lane 3), with a single stranded ISH1 specific DNA as probe. Numbers on the left show the position of DNA size markers. Arrow indicates the location of BR mRNA on the same gel (as detected by using BR cDNA as probe). (B), Southern blot hybridization ; Lane 1, Ethidium bromide staining pattern of DNA fragments produced by combined digestion of pMSb1 DNA with Tth1, Kpn1 and BamH1 ; Lane 2, hybridization pattern of lane 1 using in vivo labelled H. halobium RNA as probe. Arrow on the right indicates hybridization of RNA to a 1 kb Tth1 fragment derived from ISH1 (Fig.1C).

For most IS elements, it is not known if the ORF sequences really code for a protein in vivo. In the case of E.coli IS5, it has been recently demonstrated (17) that ORFs in this element are expressed in vivo to produce 12.3- and 37 kilo dalton proteins. For the ISH1, a stable RNA transcript complementary to the long ORF of ISH1 is shown to be present in H.halobium (see below). Thus, it is very likely that the long ORF of ISH1 codes for a polypeptide which is 270 aminoacids in length.

Evidence for the presence of ISH1 specific RNA was obtained by using two different hybridization approaches. (Fig.4A and 4B). In the first approach, uniformly 32p-labelled total RNA was obtained from H.halobium and after removal of BR mRNA from this preparation, it was used as probe against various DNA fragments which were produced by combined digestion of pMSb1 plasmid DNA with Tth1, Kpn1 and BamH1. Among the many H.halobium DNA fragments seperated by gel, only two hybridized with the in vivo labelled RNA probe and one of these (indicated with arrow in Fig.4B) corresponds to the 1.0 kbp Tth1 fragment which covers almost all of the ISH1 sequences (Fig.1C). This data clearly shows the presence of a stable RNA transcript that is complementary to some region(s) of ISH1 element. To find out if the ISH1 specific RNA is really derived from the long ORF of ISH1,

a single stranded DNA which is complementary to the coding strand of
ORF in ISH1 was used as probe against the total RNA of three different
H.halobium strains (Fig.4A). A single hybridization band was obtained
for each halobacterial RNA preparation and this clearly shows that
there is a unique-sized stable transcript which is complementary to
the ORF in ISH1. It was also interesting to see that the intensity
of the hybridization for the R1mR strain (lane 3) was significantly
greater than that of the other two strains. This finding is consistent with the data (see below) that R1mR has the highest copy number
of ISH1 element. The size of the ISH1 specific RNA is slightly longer
than that of the BR mRNA (Fig.4A, arrow) and is estimated to be around
900 nucleotides. This size is in good agreement with the size of long
ORF (810 nucleotides) in ISH1 element which is expected to code for
a polypeptide of 270 aminoacids.

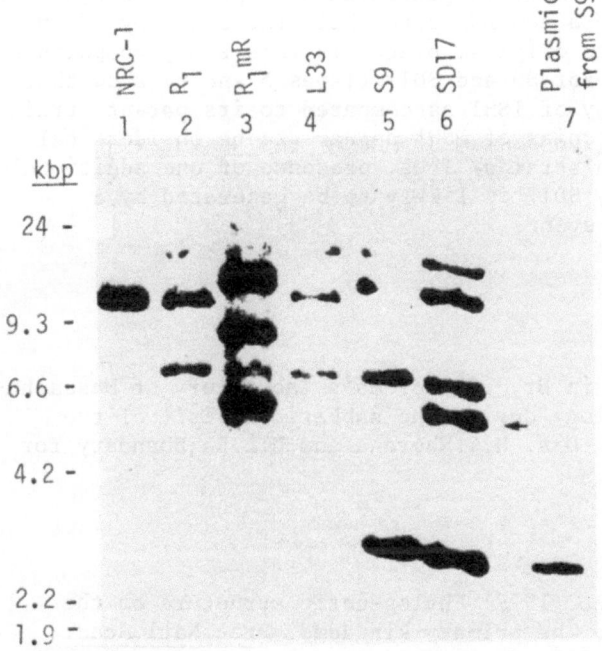

Figure 5. Southern blot
hybridization of ISH1 specific probe to Pst 1
digests of DNA from dif-
ferent strains of H.
halobium. Total DNA of
strains NRC-1, R1, R1mR,
L33, S9 and SD17 (lanes
1-6, respectively) and
plasmid DNA of S9 (lane
7), were used for the
Pst1 digestion. Numbers
on the left indicate DNA
size markers. Arrow shows
the position of a new
copy of ISH1 formed by
duplicative transposi-
tion.

ISH1 IS PRESENT IN MULTIPLE COPIES IN THE H.halobium DNA

One of the noteworthy features of transposable elements is their
presence usually in multiple copies within the genome (4). To examine
the copy number of ISH1, Pst1 digests of total DNA from six H.
halobium strains (NRC-1, R1, R1mR, L33, S9 and SD17) were analyzed
by Southern blot hybridization (18) using a probe (1.0 kbp Tth1
fragment of ISH1, Fig.1C) which is totally specific for the ISH1
sequences. Fig.5 shows that ISH1 is present from one to five or more
copies per genome depending on the strain. Wild type strain NRC-1

contains one strong band whereas strains Rl and L33 contain two
additional bands. RlmR, S9 and SD17 contain four to six bands that
hybridize with the probe. At least some of these bands are derived
from plasmid DNA of H.halobium. For example, the 3.2 kbp fragment
in digests of S9 and SD17 is also present in PstI digest of purified
plasmid DNA from strain S9 (Lanes 5,6 and 7). In addition, the relatively strong intensity of three of the hybridization bands in
RlmR (lane 3) suggests that these may also be derived from plasmid
DNA(s), because, plasmids in halobacteria are present in about four
copies (19). This could mean that strain RlmR has the largest copy
number of ISH1 sequences in its DNA.

One other interesting feature of transposable elements is
their movement (transposition) within the genome. Although, the
details of this process are not known at molecular level, various
models (20,21) have been proposed and they all indicate that a new
copy of the transposable element is generated during the transposition
event and is inserted into a new DNA site while the old copy of the
element stays behind at its original place. In this study comparison
of hybridization patterns for S9 and SD17 (lanes 5 and 6) show that
SD17 has an additional copy of ISH1 as compared to its parent strain
(S9). The remaining four copies of ISH1 appears to be on identical
size DNA fragments in both strains. Thus, presence of one additional
ISH1 in the mutant strain, SD17 is likely to be generated by a
duplicative transposition event.

ACKNOWLEDGEMENT

This work was carried out in Dr.H.G.Khorana's laboratory at Massachusetts Institute of Technology during the sabbatical visit of the
author who wishes to thank Drs. H.G.Khorana and U.L.RajBhandary for
their helpful discussions.

REFERENCES

1. Woose, C.R. and Fox,G.E. (1977) Phylogenetic structure of the
      prokaryotic domain ; The primary kingdoms. Proc.Natl.Acad.
      Sci. USA. 74, 5088-5090.
2. Bayley,S.T. and Morton,R.A. (1978) Recent developments in the
      molecular Biology of halophilic bacteria. CRC, Critical Reviews
      of Microbiology, G, 151-205.
3. Şimşek, M.,DasSarma,S., RajBhandary,U.L. and Khorana, H.G.(1982).
      A transposable element from Halobacterium halobium which inacti-
      vates the bacteriorhodopsin gene. Proc.Natl.Acad.Sci.USA. 79,
      7268-7272.
4. Calos,M.P. and Miller,J.H.(1980). Transposable elements.
      Cell. 20, 579-595.

5. Dunn,R., McCoy,J., Şimşek,M., Majumdar,A., Chang,S.H., RajBhandary, U.L. and Khorana,H.G. (1981). The bacteriorhodopsin gene. Proc.Natl.Acad.Sci.USA. 78, 6744-6748.

6. Maxam,A.M. and Gilbert,W. (1980). Sequencing end-labelled DNA with base specific chemical cleavages. Methods Enzymol. 65, 499-560.

7. Ikenaga,H. and Saigo,K. (1982). Insertion of a movable genetic element, 297, into the -T-A-T-A box for the H_3 histone gene in Drosophila melanogaster. Proc.Natl.Acad.Sci.USA 79, 4143-4147.

8. Halling,S.M. and Kleckner,N. (1982). A symmetrical six-base pair target site sequence determines Tn10 insertion specifity. Cell. 28, 155-163.

9. Lichtenstein,C. and Brenner,S. (1982). Unique Insertion site of Tn7 in the E.coli chromosome. Nature (London) 297, 601-603.

10. DasSarma,S., RajBhandary,U.L. and Khorana,H.G. (1983). Bacterio-opsin mRNA in wild type and bacterio-opsin deficient. Halobacterium halobium strains. (Proc.Natl.Acad.Sci.USA. in press.)

11. Sanger,F., Nicklen,S. and Coulson, A.R. (1977). DNA sequencing with chain termination inhibitors. Proc.Natl.Acad.Sci.USA. 74, 5463-5467.

12. Ohtsubo,H. and Ohtsuba,E. (1978) Nucleotide Sequence of an insertion element, IS1. Proc.Natl.Acad.Sci.USA. 75, 615-619.

13. Hu,S., Ohtsuba,E. and Davidson,N. (1975). Electron Microscope Heteroduplex Studies of Sequence Relation Among Plasmids of Escherichia coli : Structure of F13 and Related F-primes. J.Bacteriol. 122, 749-763.

14. Krŏger,M. and Hobom,G. (1982). Structural analysis of insertion sequence IS5. Nature (London) 297, 159-162.

15. DasSarma,S., RajBhandary,U.L. and Khorana,H.G. (1983). High frequency spontaneous mutations in the bacterio-opsin gene in Halobacterium halobium is mediated by transposable elements. Proc.Natl.Acad.Sci.USA. 80, 2201-2205.

16. Xu,W.L. and Doolittle,U.F. (1983) Structure of the archaebacterial transposable element, ISH50. Nucleic Acids Res. 11, 4195-4199.

17. Rak,B., Lusky, M. and Hable,M. (1982). Expression of two proteins from overlapping and oppositely oriented genes on transposable DNA insertion element, IS5. Nature (London) 297, 124-128.

18. Southern,E.M. (1975) Detection of specific sequences among DNA fragments separated by gel electrophoresis. J.Mol.Biol.98, 503-517.

19. Weidinger,G., Klotz,G. and Goebel,W. (1977) A large plasmid from Halobacterium halobium carrying genetic information for gas vacuole formation. Plasmid 2, 377-380.

20. Botstein,D, and Kleckner,N. (1977). In DNA Insertion Elements, plasmids and Episomes. eds. Bukhari, A.I., Shapiro,J.A. and Adhya,S.L. (New York: Cold Spring Harbor Laboratory). p.185-203.

21. Shapiro,J.A. (1979) Molecular model for the transposition and replication of bacteriophage Mu and other transposable elements. Proc.Natl.Acad.Sci.USA. 76, 1933-1937.

STREPTOMYCETE PLASMID CLONING VECTORS

Charles J. Thompson and Julian E. Davies

Biogen S.A.
46, route des Acacias
1227 Geneva/Switzerland

Academic and industrial interests have recently motivated a rapid development of streptomycete recombinant DNA technology (9, 15, 25). This was made possible by the availability of plasmids and the critical discovery by Bibb et al. that plasmid DNA could be introduced to streptomycete protoplasts following polyethylene–glycol treatment (1). Since then, interspecific genomic cloning experiments have allowed the isolation of streptomycete antibiotic resistance genes which are normally associated with the production of antibiotics and the tyrosinase gene which catalyzes melanine synthesis (2, 13, 16, 20, 23). These genes have been used to construct vectors from plasmid and bacteriophage replicons (5). Here we will focus primarily on the development of the vectors most commonly used today which were derived from the streptomycete plasmids SLP1 and pIJ101.

The SLP1 replicon, which originated as part of the S. coelicolor chromosome, is found in a series of self-transmissible plasmids (3). The SLP1 sequence will excise and transfer to S. lividans where it replicates autonomously with a copy number of 3-5 relative to the chromosome. It seems that the SLP1 plasmids have common requirements for maintenance in S. lividans and therefore contain a homologous 'core' region. Since excision is imprecise, these plasmids also have a heterogeneous region which represents various end-point fusions. Thus, the largest member of the family, SLP1.2 (14.5 kb, fig. 1a), contains a dispensible region which has the only cleavage sites for Bam H1, Pst 1, Sph 1, Xho 1, and Kpn 1 (3, 21). The implication that DNA could be inserted here without destroying essential plasmid functions was confirmed in the first streptomycete cloning experiments described below.

The 8.9 kb pIJ101 plasmid which became available several years

19

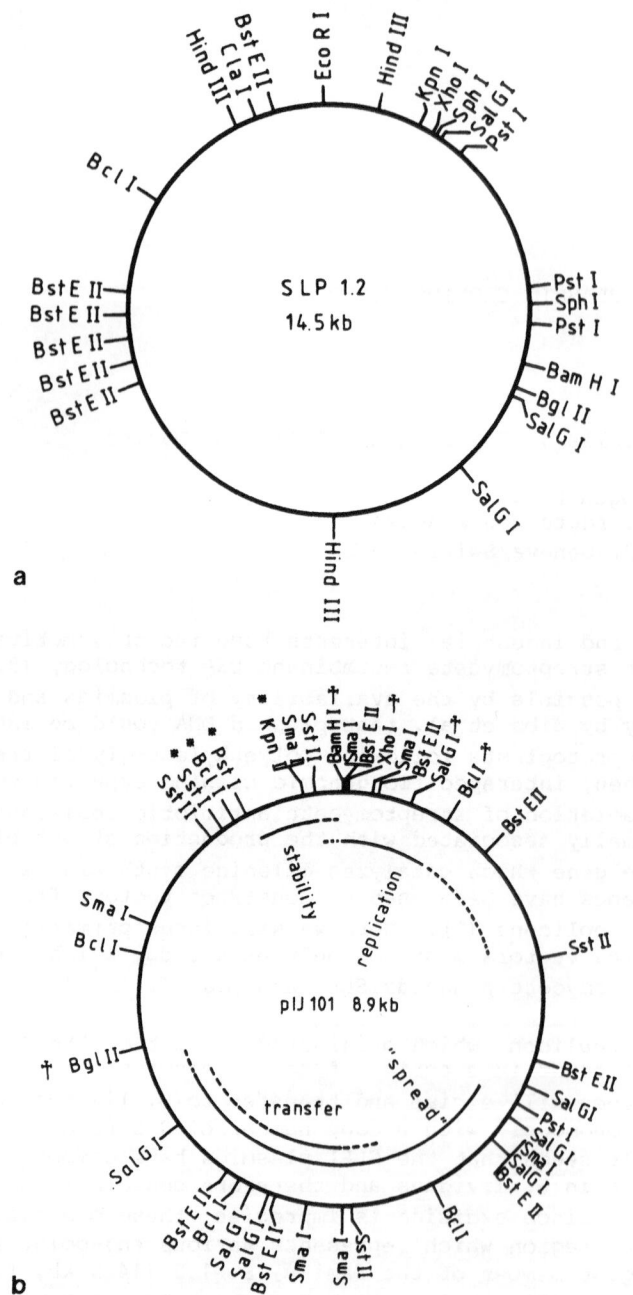

Figure 1. Streptomycete plasmids used to construct cloning vectors:
a. SLP1.2 (21) and b. pIJ101 (14).

later, has also provided a replicon whose properties are convenient for vector construction (13, 14, 18; fig. 1b). All of the functions which are needed for replication are localized in a small (2.1 kb) Sst II fragment (14). Transfer genes, which are also clustered, can be inactivated for biocontainment purposes or reinserted (using restriction enzymes or in vivo recombination) for convenient screening of cloned functions in different strains or mutants. The high copy number of pIJ101 (40-300 copies per chromosome) makes it easy to isolate large amounts of plasmid DNA and allows for amplified expression of cloned genes. Its host range is broad (these plasmids transform approximately 80% of the Streptomyces sp. tested) and may be limited primarily by restriction systems which are found in many species.

In order to use plasmids as vectors, one needs to be able to select for or at least detect their presence in the cell. Fortunately, most streptomycete plasmids code for an easily detectable phenotype called lethal zygosis (ltz; 1, 3, 14, 17). The transfer of an ltz+ plasmid into a plasmid free strain inhibits the growth of the recipient. If a spore of a ltz+ strain is plated out on a lawn of a plasmid free strain, it forms a zone of inhibition which often looks like a bacteriophage plaque and has been called a pock. There are many types of pock morphology which vary with respect to their size and sporulation patterns. Although the biological significance of this phenomenonon is not understood, it has served as a valuable genetic marker to classify and detect plasmids.

In many cases, however, it is more convenient to be able to use a marker such as an antibiotic resistance gene to select directly for the presence of a plasmid. Antibiotic producing Streptomycetes sp. contain genes which protect them from the toxic effects of their products (22). Some of these genes which code for enzymes that modify the target site or inactivate the antibiotic were expected to confer easily selectable, dominant resistance phenotypes (see Table 1). They were therefore chosen to demonstrate the practicality of interspecific cloning in streptomycetes and to provide genes for vector construction. Initially, the genes for methylenomycin, thiostrepton, and neomycin resistance were cloned from S. coelicolor, S. azureus, and S. fradiae respectively into S. lividans using SLP1.2 as vector. Subsequently, genes which code for resistance to erythromycin, viomycin, hygromycin, streptomycin, kanamycin, novobiocin, destomycin, ribostamycin, and racemomycin have been isolated by interspecific cloning as summarized in Table 1. These antibiotic resistance genes have made it possible to construct a variety of selectable streptomycete vectors.

In addition to having genes which allow selection for the presence of the plasmid, it is also useful to have a marker with an easily detectable product. This could allow the identification of plasmids containing inserts which inactivate that gene. The tyrosinase gene, originally cloned from S. antibioticus genomic DNA

Table 1. Cloned streptomycete antibiotic resistance genes

Resistance conferred	Enzyme specified	Source of gene	Reference
thiostrepton	23S rRNA methylase	S. azureus	(20, 22)
neomycin	phosphotransferase	S. fradiae	(20, 22, 24)
neomycin	acetyltransferase	S. fradiae	(20, 22)
hygromycin	phosphotransferase	S. hygroscopicus	(12)
streptomycin	phosphotransferase	S. glaucesens	(8)
kanamycin	acetyltransferase	S. kanamyceticus	(16)
kanamycin	?	S. kanamyceticus	(16)
destomycin	?	S. rimofacirns	(16)
ribostamycin	phosphotransferase	S. ribosidificus	(16)
racemomycin	?	S. lavendulae	(16)
erythromycin	23S rRNA methytlase	S. erythreus	(21, 22)
viomycin	phosphotransferase	S. vinaceus	(22, 23)
viomycin	?	S. vinaceus	(23)
methylenomycin	?	S. coelicolor	(2)

into S. lividans on a vector derived from SLP1.2, has been useful in this regard (13). The gene product, tyrosinase (EC 1.14.18.1), catalyzes the enzymatic conversion of tyrosine to melanine. Since S. lividans (as well as the majority of streptomycetes) cannot make this conversion, strains containing the recombinant plasmid were detected as a black S. lividans colony.

These potential vector components had to be combined in a way which would generate a vector of minimal size having a selectable gene and unique restriction endonuclease sites for the insertion of DNA. Ideally, the cloning sites should lie within a gene whose regulation and product are understood. Essential regions for pIJ101 functions (i.e. replication, stability, and transfer) and antibiotic resistance were established at the same time by subcloning resistance genes into various sites of the plasmid (14; fig. 2b). SLP1.2 has not yet been as exhaustively studied. Although many potential vectors were generated in these experiments, only the most widely used ones, pIJ61 and pIJ702, will be compared here; details of their construction have been presented elsewhere (13, 14, 21).

pIJ61 was made by incorporating a neomycin phosphotransferase gene (aph) and thiostrepton resistance gene (tsr) into a deletion derivative of SLP1.2 (21; fig. 2a). tsr is a useful gene for selective purposes; most Streptomycetes sp. are sensitive to thiostrepton and resistant mutants rarely occur. Since the aph structural gene provides unique BamH1 and Pst1 sites in pIJ61, restriction fragments can be readily inserted here and identified by insertional inactivation of the neomycin resistance gene. The exression of the cloned gene could

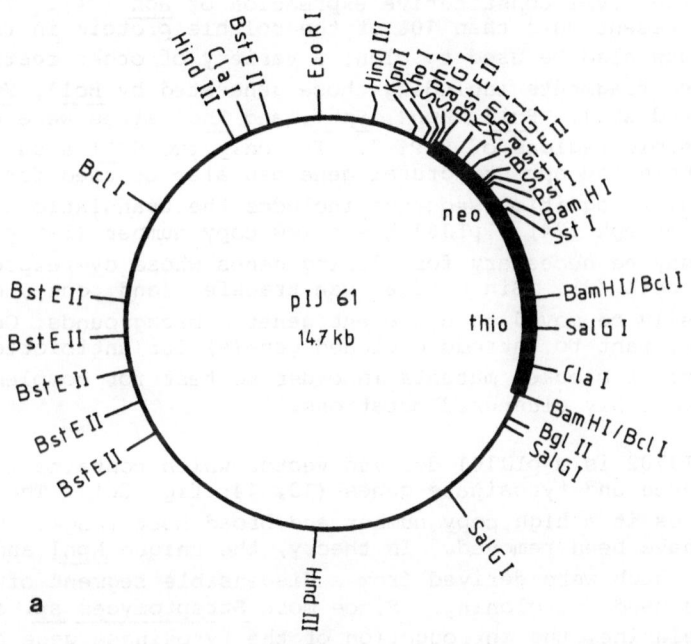

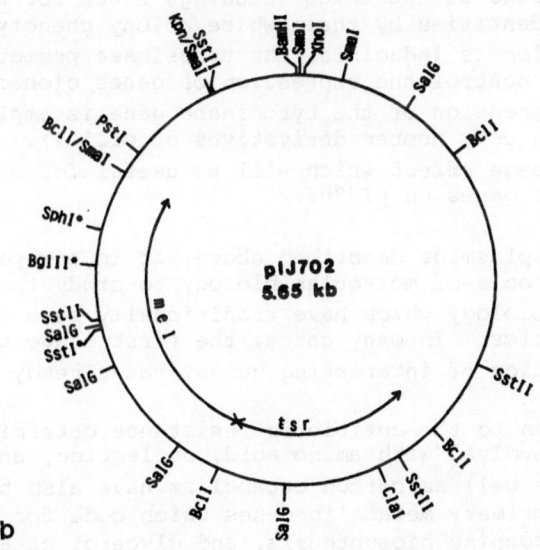

Figure 2. Streptomycete plasmid cloning vectors: a. pIJ61 (21) and b. pIJ702 (13).

be directed by the aph promoter (21, 24). This promoter is associated with high level constitutive expression of aph (24). Its gene product can represent more than 10% of the soluble protein in the cell (24). pIJ61 can also be used to clone a variety of other restriction endonuclease fragments including those generated by Bcl1, Kpn1, Sph1, Sst1, and Xba1. Bcl1, Kpn1, Sph1, and Xho1 sites were derived from dispensible regions of SLP1.2. The only two Sst1 sites of pIJ61 which are within the aph structural gene can also be used for cloning (24). The Xba1 recognition sequence includes the translational termination signal of aph (24). pIJ61 has a low copy number (3-5 per chromosome) which may be necessary for cloning genes whose overexpression is toxic to the cell (2). Since pIJ61 can transfer (and pock), cloned genes can easily be moved to different genetic backgrounds. One might, for example, want to introduce cloned gene(s) for antibiotic production to a series of blocked mutants in order to test for complementation of other possibly clustered mutations.

pIJ702 is a pIJ101 derived vector which contains the thiostrepton resistance and tyrosinase genes (13, 14; fig. 2b). The pIJ101 replicon gives it a high copy number and broad host range. Transfer functions have been removed. In theory, the unique Kpn1 and Pst1 sites of pIJ702 which were derived from a dispensible segment of pIJ101 could also be used for cloning. Since most Streptomyces sp. cannot synthesize melanine, the introduction of the tyrosinase gene on pIJ702 results in the production of black colonies. Plasmids containing fragments inserted at the unique cleavage sites for Bgl II, Sph 1, or Sst 1 can be identified by their white colony phenotype. Since tyrosinase expression is inducible, the tyrosinase promoter might be used to allow or to control the expression of genes cloned into these sites. The expression of the tyrosinase gene is amplified on pIJ702 (and other high copy number derivatives of pIJ101). This is presumably a gene dosage effect which will be useful for overexpressing other desirable genes on pIJ702.

Using the plasmids described above, it is now possible to use the sophisticated tools of molecular biology to study the problems of streptomycete biology which have traditionally been studied using classical genetics. In many cases, the first stage of these studies, i.e. the isolation of interesting genes, has already been achieved.

In addition to the antibiotic resistance determinants mentioned above, genes involved with amino acid, nucleotide, and antibiotic biosynthesis as well as carbon catabolism have also been cloned. These include primary metabolic genes which code for histidine, arginine, and guanine biosynthesis, and glycerol catabolism (glycerol kinase and dehydrogenase) (19, 23). Structural and regulatory genes of secondary metabolism have also been isolated. These include the structural gene for an o-methyltransferase which is needed for biosynthesis of the antibiotic prodigiosin and genes involved in the regulation of candicidin and streptomycin biosynthesis (7, 11).

24

Studies of the mechanisms by which streptomycetes control the expression of these genes are currently in progress. Although we now know the complete nucleotide sequence of aph and its surrounding regions, regulatory sequences have not yet been identified experimentally (24). Although sequences analogous to the termination signals for transcription and translation of other bacterial messages are present, initiation signals cannot be found. Promoter probe vectors have been used to isolate sequences with transcription initiation activity, however no consensus promoter sequence has emerged from these studies (4). Streptomycete molecular biology is still in a primative state in this respect but is currently being intensively studied.

Complementary to the application of gene cloning to the area of basic streptomycete molecular biology, we can expect that this new technology and information will be applied to many industrial problems (9,15, 25). Finding new ways to discover or increase the production of antibiotics using recombinant DNA technology has been the subject of much speculation. Streptomyces also have some potentially useful characteristics for the production of pharmaceutically important proteins. They are well understood microorganisms from a fermentation point of view with no pathogenicity associated with their use.

REFERENCES

1. M. J. Bibb, J. M. Ward, and D. A. Hopwood. 1978. Transformation of plasmid DNA into streptomycetes at high frequency. Nature (London) 274:398-400.
2. M. J. Bibb, J. L. Schottel and S. N. Cohen. 1980. A DNA cloning system for interspecies gene transfer in antibiotic producing Streptomyces. Nature (London) 284:526-531.
3. M. J. Bibb, J. M. Ward, T. Kieser, S. N. Cohen, and D. A. Hopwood. 1981. Excision of chromosomal DNA sequences from Streptomyces coelicolor forms a novel family of plasmids detectable in Streptomyces lividans. Mol. Gen. Genetics 184:230-240.
4. M. J. Bibb, and S. N. Cohen. 1982. Gene expression in Streptomyces: construction and application of promoter-probe plasmid vectors in Streptomyces lividans. Mol. Gen. Genetics 187:265-277.
5. K. F. Chater, D. A. Hopwood, T. Kieser, and C. J. Thompson. 1982. Gene cloning in Streptomyces. Curr. Top. Microbiol. Immunol. 96:69-75.
6. K. F. Chater, and C. J. Bruton. Mutational cloning in Streptomyces and the isolation of antibiotic production genes. Gene: (in press).
7. J. S. Feitelson, and D. A. Hopwood. 1983. Cloning of a Streptomyces gene for an o-methyltransferase involved in antibiotic biosynthesis. Mol. Gen. Genetics 190:394-398.
8. G. Hintermann, personal communication.
9. D. A. Hopwood, M. J. Bibb, C. J. Bruton, K. F. Chater, J. S.

Feitelson and J. A. Gil. 1983. Cloning Streptomyces genes for antibiotic production. Trends in Biotechnology 1:42-48.

10. D. A. Hopwood, T. Kieser, H. M. Wright, and M. J. Bibb. 1983. Plasmids, recombination and chromosome mapping in Streptomyces lividans 66. J. Gen. Microbiol. 129:2257-2269.

11. S. Horinouchi, O. Hara, and T. Beppu. 1983. Cloning of a pleiotropic gene that positively controls biosynthesis of A-factor, actinorhodin, and prodigiosin in Streptomyces coelicolor A3(2) and Streptomyces lividans. J. Bacteriol. 155:1238-1248.

12. A. Jimenez, personal communication.

13. E. Katz, C. J. Thompson, and D. A. Hopwood. 1983. Cloning and expression of the tyrosinase gene from Streptomyces antibioticus in Streptomyces lividans. J. Gen. Microbiol. 129:2703-2714.

14. T. Kieser, D. A. Hopwood, H. M. Wright, and C. J. Thompson. 1982. pIJ101, a multi-copy broad host-range Streptomyces plasmid: functional analysis and development of DNA cloning vectors. Mol. Gen. Genetics 185:223-238.

15. J. F. Martin, and J. A. Gil. 1984. Cloning and expression of antibiotic production genes. Bio/Technol. 1:63-72.

16. T. Murakami, C. Nojiri, H. Toyama, E. Hayashi, K. Katumata, H. Anzai, Y. Matsuhashi, Y. Yamada, and K. Nagaoka. 1983. Cloning of antibiotic-resistance genes in Streptomyces. J. Antibiotics 36:1305-1311.

17. T. Murakami, C. Nojiri, H. Toyama, E. Hayashi, and K. Nagaoka. 1983. Pock forming plasmids from antibiotic producing Streptomyces. J. Antibiotics 36:429-434.

18. J. L. Pernodet, and M. Guerineau. 1981. Isolation and characterization of streptomycete plasmids. Mol. Gen. Genet. 182:53-59.

19. E. T. Seno, C. J. Bruton. 1984. The glycerol utilization operon of Streptomyces coelicolor: Genetic mapping of gyl mutations and the analysis of cloned gly DNA. Mol. Gen. Genetics: 193:119-128.

20. C. J. Thompson, J. M. Ward, and D. A. Hopwood. 1980. DNA cloning in Streptomyces: resistance genes from antibiotic producing species. Nature (London) 286:525-527.

21. C. J. Thompson, T. Kieser, and D. A. Hopwood. 1982. Physical analysis of antibiotic-resistance genes from Streptomyces and their use in vector construction. Gene 20:51-62.

22. C. J. Thompson, R. H. Skinner, J. Thompson, J. M. Ward, D. A. Hopwood, and E. Cundliffe. 1982. Biochemical characterization of resistance determinants cloned from antibiotic-producing streptomycetes. J. Bacteriol. 151:678-685.

23. C. J. Thompson, J. M. Ward, and D. A. Hopwood. 1982. Cloning of antibiotic resistance and nutritional genes in streptomycetes. J. Bacteriol. 151:668-677.

24. C. J. Thompson, and G. S. Gray. 1983. Nucleotide sequence of a streptomycete aminoglycoside phosphotransferase gene and its relationship to phosphotransferases encoded by resistance plasmids. Proc. Natl. Acad. Sci. 80:5190-5194.

25. C. J. Thompson, and J. E. Davies. 1984. Genetic engineering and antibiotics. Trends in biotechnology:in press.

GENETIC INSTABILITY IN STREPTOMYCETES

Ralf Hütter and Gilberto Hintermann

Mikrobiologisches Institut
Eidgenössische Technische Hochschule
CH-8092 Zürich, Switzerland

INTRODUCTION

Streptomycetes and other actinomycetes can exhibit a remarkable degree of phenotypic variability, which is frequently due to genetic variability. This behaviour attracted interest, because it affects not only morphological characters but also various enzymatic activities, antibiotic resistance and the commercially important antibiotic production. The problem has already been dealt with extensively in various review articles. Therefore only a short summary with a number of recent relevant references will be given, focusing on the role of plasmids and on chromosomal instability and without any claim of completeness.

Remarkably "instability conditions" usually affect various phenotypic traits in parallel. This led to the assumption that pleiotropic mutations have occurred (e.g. Shaw and Piwowarski, 1977; Suter et al., 1978; Redshaw et al., 1979; Nakano and Ogawara, 1980; Furumai et al., 1982). But in cases where more detailed analyses were done it was found, that separate mutational events, occurring in parallel, were most likely responsible for the various defects (e.g. Hütter et al., 1981; Crameri et al., 1983; Davies, 1983; Schrempf, 1983b).

INVOLVEMENT OF PLASMIDS

Plasmids occur abundantly in streptomycetes (see Okanishi, 1979; Hayakawa et al., 1979a; Hopwood et al., 1979; Schrempf, 1979, 1981a; Okanishi et al., 1980; Omura et al., 1981; Hershberger, 1982; Kirby et al., 1982; Toyama et al., 1982; Davies, 1983), and possibly in other actinomycetes as well (e.g. Kasweck et al., 1982). Plasmids may also arise by excision of chromosomal DNA sequences during crosses between plasmid free strains (e.g. Bibb et al., 1981). While most plasmids are circular, some linear plasmids have also been described (Hayakawa et al., 1979b; Hirochika and Sakaguchi, 1982).

For most cases the biological role of plasmids is not known. But some have clearly been established as fertility factors, containing information for "pock" formation ("lethal zygosis"), plasmid transfer and sometimes chromosome mobilization (see Vivian, 1971; Hopwood et al., 1973; Hopwood, 1983a; Hopwood and Wright, 1973; Schrempf et al., 1975; Bibb et al., 1977; Friend et al., 1978; Bibb and Hopwood, 1981; Ogata et al., 1982; Murakami et al., 1983). In the field of antibiotic biosynthesis and naturally occurring antibiotic resistance, only the structural genes for methylenomycin formation and resistance were clearly shown to be plasmid coded (Kirby and Hopwood, 1977; Aguilar and Hopwood, 1982), while for other cases the evidence is at most circumstantial (e.g. Hopwood, 1978, 1983b; Freeman and Hopwood, 1978; Okanishi and Umezawa, 1978; Okanishi, 1979; Ochi and Katz, 1980; Baltz et al., 1981; Schrempf, 1981a, 1981b; Sermonti and Lanfaloni, 1982; Ikeda et al., 1982; Davies, 1983). And in other cases a clear chromosomal localization was established (e.g. Wright and Hopwood, 1976; Rudd and Hopwodd, 1979, 1980; Okanishi, 1981; Ahmed and Vining, 1983; Hopwood, 1983b). Also for various other mutational types, e.g. loss of aerial mycelium formation, loss of melanin formation, arginine auxotrophy, plasmid involvement has been postulated, but again without thorough proof (e.g. Fedorenko and Danilenko, 1980; Nakano et al., 1980).

On the contrary, in several cases where sound proof was searched for plasmid localization of structural genes, the mutational events were finally found to be chromosomal (see below). In these cases a "regulatory role" was claimed for the plasmids, i.e. for certain plasmid functions (see e.g. Ikeda et al., 1982; Davies, 1983; Hopwood, 1983b).

Plasmids seem sometimes to be involved in the formation of phage-like particles (e.g. Ogata et al., 1982) or have been found to be themselves prophages (e.g. Chung, 1982).

Under conditions leading to genetic instability, plasmids can sometimes be lost, but frequently seem to undergo rearrangements, accompanied by deletion of part of the plasmid DNA (Komatsu et al.,

1981; Schrempf, 1981a, 1982, 1983a, 1983b). Due to the fact, that in most cases the strains have been subject to harsh mutagenic treatments, parallel or consecutive occurrence of mutational events involving plasmid DNA (where present) as well as chromosomal DNA (see below) may be the rule.

CHROMOSOMAL INSTABILITY

Besides the above mentioned shifts in plasmid content or plasmid structure, various chromosomal changes have been described. Such changes were observed by conventional genetic analyses of crosses, by one dimensional agarose gel electrophoresis, by the "double label method" (Hintermann et al., 1982) and in hybridisation experiments with cloned genes (i.e. DNA areas) of unstable chromosomal sections.

Conventional genetic analyses have shown, that "unstable muta-tions" behave as chromosomal lesions. In some instances it was sug-gested that transposition events occur (Sermonti et al., 1978, 1980; Puzynina et al., 1979), but the available data also permit alterna-tive explanations (see e.g. Lanfaloni et al., 1980). The question then arose which molecular events could be observed after mutations in unstable regions have occurred. The present data suggest two main mechanisms: deletion formation and amplification of DNA stret-ches.

In the case of melanin formation it has been shown, that many melanin negative strains have lost the structural gene for tyrosi-nase formation (deletion >10 kb; Schrempf 1982, 1983a, 1983b; G. Hintermann, unpublished data). Similarly it was found that strepto-mycin supersensitive mutants (strS⁻) of Streptomyces glaucescens, which are defective in streptomycin phosphotransferase (Ono et al., 1983), also carry a deletion of >10 kb (G. Hintermann, unpublished data).

In addition to large deletions, amplification of certain stretches of chromosomal DNA have been observed (Robinson et al., 1981; Ono et al., 1982; Schrempf, 1982, 1983a, 1983b; Fishman and Hershberger, 1983; Orlova and Danilenko, 1983). The size of the amplified DNA stretches varies from about 2 to about 35 kb (M. Hasegawa, unpub-lished data from this laboratory). The molecular events leading to the amplifications, the structure of the amplified DNA sections and their significance are not yet understood. It may be significant that in all cases where the analysis was done, mutant strains car-rying amplifications also carried deletions.

REFERENCES

Aguilar, A., and Hopwood, D.A., 1982, Determination of methylenomycin A synthesis by the pSV1 plasmid from Streptomyces violaceus-ruber SANK-95570, J.Gen.Microbiol. 128:1893.

Ahmed, Z.U., and Vining, L.C., 1983, Evidence for a chromosomal location of the genes coding for chloramphenicol production in Streptomyces venezuelae, J.Bacteriol. 154:239.

Baltz, R.H., Seno, E.T., Stonesifer, J., P. Matsushima, and Wild, G.M., 1981, Genetics and biochemistry of tylosin production by Streptomyces fradiae, in: "Microbiology 1981", D. Schlessinger, ed., American Society for Microbiology, Washington DC.

Bibb, M.J., Freeman, R.F., and Hopwood, D.A., 1977, Physical and genetical characterization of a second sex factor, SCP2, for Streptomyces coelicolor A3(2), Mol.Gen.Genet. 154:155.

Bibb, M.J., and Hopwood, D.A., 1981, Genetic studies of the fertility plasmid SCP2 and its SCP2* variants in Streptomyces coelicolor A3(2), J.Gen.Microbiol. 126:427.

Bibb, M.J., Ward, J.M., Kieser, T., Cohen, S.N., and Hopwood, D.A., 1981, Excision of chromosomal DNA sequences from Streptomyces coelicolor forms a novel family of plasmids detectable in Streptomyces lividans, Mol.Gen.Genet. 184:230.

Chung, S.-T., 1982, Isolation and characterization of Streptomyces fradiae plasmids which are prophage of actinophage ØSF1, Gene 17:239.

Crameri, R., Kieser, T., Ono, H., Sanchez, J., and Hütter, R., 1983, Chromosomal instability in Streptomyces glaucescens - mapping of streptomycin-sensitive mutants, J.Gen.Microbiol. 129:519.

Davies, J.E., 1983, Plasmids in antibiotic-producing organisms: their possible role in biosynthesis and resistance, in: "Proc. of the IVth Internat.Symp. on Genetics of Industrial Microorganisms (GIM 82)", Y. Ikeda, and T. Beppu, eds., Kodansha Ltd., Tokyo.

Fedorenko, V.A., and Danilenko, V.N., 1980, Instability of natural multiple drug resistance in actinomycetes, Antibiotiki 25:170.

Fishman, S.E., and Hershberger, C.L., 1983, Amplified DNA in Streptomyces fradiae, J.Bacteriol. 155:459.

Freeman, R.F., and Hopwood, D.A., 1978, Unstable naturally occurring resistance to antibiotics in Streptomyces, J.Gen.Microbiol. 106:377.

Friend, J., Warren, M., and Hopwood, D.A., 1978, Genetic evidence for a plasmid controlling fertility in an industrial strain of Streptomyces rimosus, J.Gen.Microbiol. 106:201.

Furumai, T., Takeda, K., and Okanishi, M., 1982, Function of plasmids in the production of aureothricin, I. Elimination of plasmids and alteration of phenotypes caused by protoplast regeneration in Streptomyces kasugaensis, J.Antibiotics 35:1367.

Hayakawa, T., Otake, N., Yonehara, H., Tanaka, T., and Sakaguchi, K., 1979a, Isolation and characterization of plasmids from Streptomyces, J.Antibiotics 32:1348.

Hayakawa, T., Tanaka, T., Sakaguchi, K., Otake, N., and Yonehara, H., 1979b, A linear plasmid-like DNA in Streptomyces sp. producing lankacidin group antibiotics, J.Gen.Microbiol. 25:255.

Hershberger, C.L., 1982, Recombinant DNA systems for application to antibiotic fermentation in Streptomyces, Ann.Rept.Ferm.Proc. 5:101.

Hintermann, G., Crameri, R., Kieser, T., and Hütter, R., 1982, A technique for the detection of large DNA alterations in complex genomes, Anal.Biochem. 121:327.

Hirochika, H., and Sakaguchi, K., 1982, Analysis of linear plasmids isolated from Streptomyces: association of protein with the ends of the plasmid DNA, Plasmid 7:59.

Hopwood, D.A., 1978, Extrachromosomal determined antibiotic production, Ann.Rev.Microbiol. 32:373.

Hopwood, D.A., 1983a, Genetic manipulation in Streptomyces, in: "Proc.of the IVth Internat.Symp. on Genetics of Industrial Microorganisms (GIM 82)", Y. Ikeda, and T. Beppu, eds., Kodansha Ltd., Tokyo.

Hopwood, D.A., 1983b, Actinomycete genetics and antibiotic production, in: "Biochemistry and genetic regulation of commercially important antibiotics", L.C. Vining, ed., Addison-Wesley Publ. Co., London.

Hopwood, D.A., Chater, K.F., Dowding, J.E., and Vivian, A., 1973, Advances in Streptomyces coelicolor genetics, Bacteriol.Rev. 37:371.

Hopwood, D.A. and Wright, H.W., 1973, Transfer of a plasmid between Streptomyces species, J.Gen.Microbiol. 77:187.

Hopwood, D.A., Bibb, M.J., Ward, J.M., and Westpheling, J., 1979, Plasmids in Streptomyces coelicolor and related species, in: "Plasmids of medical, environmental and commercial importance", K.N. Timmis, and A. Pühler, eds., Elsevier/North Holland, Amsterdam.

Hütter, R., Kieser, T., Crameri, R., and Hintermann, G., 1981, Chromosomal instability in Streptomyces glaucescens, Zbl.Bakt.Suppl. 11:551.

Ikeda, H., Tanaka, H., and Omura, S., 1982, Genetic and biochemical features of spiramycin biosynthesis in Streptomyces ambofaciens - curing, protoplast regeneration and plasmid transfer, J.Antibiotics 35:507.

Kasweck, K.L., Little, M.L., and Bradley, S.G., 1982, Plasmids in mating strains of Nocardia asteroides, Dev.Industr.Microbiol. 23:279.

Kirby, R., and Hopwood, D.A., 1977, Genetic determination of methyle-
    nomycin synthesis by the SCP1 plasmid of Streptomyces coelico-
    lor A3(2), J.Gen.Microbiol. 98:239.
Kirby, R., Lewis, E., and Botha, C., 1982, A survey of Streptomyces
    species for covalently closed circular (CCC) DNA using a varie-
    ty of methods, FEMS Microbiol.Lett. 13:79.
Komatsu, K., Leboul, J., Harford, S., and Davies, J., 1981, Studies
    of plasmids in neomycin-producing Streptomyces fradiae, in:
    "Microbiology 1981", D. Schlessinger, ed., American Society
    for Microbiology, Washington DC.
Lanfaloni, L., Micheli, M.R., and Sermonti, G., 1980, "Salto" del
    transposone SCTn1 su un plasmide in Streptomyces coelicolor,
    Riv.Biol. 73:431.
Murakami, T., Nojiri, C., Toyama, H., Hayashi, E., Yamada, Y., and
    Nagaoka, K., 1983, Pock forming plasmids from antibiotic-pro-
    ducing Streptomyces, J.Antibiotics 36:429.
Nakano, M.M., Ogawara, H., 1980, Multiple effects induced by unsta-
    ble mutation in Streptomyces lavendulae, J.Antibiotics 33:420.
Nakano, M.M., Ozawa, K., and Ogawara, H., 1980, Possible involve-
    ment of a plasmid in arginine auxotrophic mutation in Strepto-
    myces kasugaensis, J.Bacteriol. 143:1501.
Ochi, K., and Katz, E., 1980, Genetic analysis of the actinomycin-
    producing determinants (plasmid) in Streptomyces parvulus using
    the protoplast fusion technique, Canad.J.Microbiol. 26:1460.
Ogata, S., Suenaga, H., and Hayashida, S., 1982, Pock formation of
    Streptomyces endus with production of phage taillike particles,
    Appl.Environm.Microbiol. 43:1182.
Okanishi, M., 1979, Plasmids and antibiotic synthesis in streptomy-
    cetes, in: "Proc.3d Internat.Symp. on Genetics of Industrial
    Microorganisms (GIM 78)", O.K. Sebek, and A.I. Laskin, eds.,
    American Society for Microbiology, Washington DC.
Okanishi, M., 1981, Role of plasmid genes in aureothricin production,
    in: "Advances in Biotechnology Vol. III", M. Moo-Young, C. Vezina,
    and K. Singh, eds., Pergamon Press, Toronto.
Okanishi, M., Manome, T., and Umezawa, H., 1980, Isolation and cha-
    racterization of plasmid DNAs in actinomycetes, J.Antibiotics
    33:88.
Okanishi, M., and Umezawa, H., 1978, Plasmids involved in antibiotic
    production in streptomycetes, in: "Genetics of Actinomycetales,
    Proc.Internat.Coll. Borstel 1976", E. Freerksen, I. Tarnok, and
    J.H. Thumin, eds., Gustav Fischer Verlag, Stuttgart.
Omura, S., Ikeda, H., and Tanaka, H., 1981, Extraction and charac-
    terization of plasmids from macrolide antibiotic-producing
    streptomycetes, J.Antibiotics 34:478.
Ono, H., Crameri, R., Hintermann, G., and Hütter, R., 1983, Hydroxy-
    streptomycin production and resistance in Streptomyces glau-
    cescens, J.Gen.Microbiol. 129:529.

32

Ono, H., Hintermann, G., Crameri, R., Wallis, G., and Hütter, R., 1982, Reiterated DNA sequences in a mutant strain of Streptomyces glaucescens and cloning of the sequence in Escherichia coli, Mol.Gen.Genet. 186:106.

Orlova, V.A., and Danilenko, V.N., 1983, Multiplication of DNA fragment in Streptomyces antibioticus producing oleandomycin, Antibiotiki 28:163.

Puzynina, G.G., Danilenko, V.N., Vasil'chenko, L.G., Mkrtumian, N.M., and Lomovskaya, N.D., 1979, Determination of the erythromycin resistance in Streptomyces coelicolor A3(2), Genetika 15:1511 (Engl.transl.)

Redshaw, P.A., McCann, P.A., Pentella, M.A., and Pogell, B.M., 1979, Simultaneous loss of multiple differentiated functions in aerial mycelium-negative isolates of streptomycetes, J.Bacteriol. 137: 891.

Robinson, M., Lewis, E., and Napier, E., 1981, Occurrence of reiterated DNA sequences in strains of Streptomyces produced by an interspecific protoplast fusion, Mol.Gen.Genet. 182:336.

Rudd, B.A.M., and Hopwood, D.A., 1979, Genetics of actinorhodin biosynthesis by Streptomyces coelicolor A3(2), J.Gen.Microbiol. 114:35.

Rudd, B.A.M., and Hopwood, D.A., 1980, A pigmented mycelial antibiotic in Streptomyces coelicolor: control by a chromosomal gene cluster, J.Gen.Microbiol. 119:333.

Schrempf, H., 1979, Plasmids from streptomycetes, in: "Plasmids of medical, environmental and commercial importance", K.N. Timmis, and A. Pühler, eds., Elsevier/North Holland, Amsterdam.

Schrempf, H., 1981a, Function of plasmid genes in streptomycetes, Zbl.Bakt.(Suppl.) 11:545.

Schrempf, H., 1981b, Function of plasmids in producers of macrolides, in: "Advances in Biotechnology Vol. III", M. Moo-Young, C. Vezina, and K. Singh, eds., Pergamon Press, Toronto.

Schrempf, H., 1982, Plasmid loss and changes within the chromosomal DNA of Streptomyces reticuli, J.Bacteriol. 151:701.

Schrempf, H., 1983a, Genetic intability in Streptomyces, in: "Proc. of the IVth Internat.Symp. on Genetics of Industrial Microorganisms (GIM 82)", Y. Ikeda, and T. Beppu, eds., Kodansha Ltd., Tokyo.

Schrempf, H., 1983b, Deletion and amplification of DNA sequences in melanin-negative variants of Streptomyces reticuli, Mol.Gen. Genet. 189:501.

Schrempf, H., Bujard, H., Hopwood, D.A., and Goebel, W., 1975, Isolation of covalently closed circular deoxyribonucleic acid from Streptomyces coelicolor A3(2), J.Bacteriol. 121:416.

Sermonti, G., Lanfaloni, L., and Micheli, M.R., 1980, A jumping gene in Streptomyces coelicolor A3(2), Molec.Gen.Genet. 177:453.

Sermonti, G., and Lanfaloni, L., 1982, Antibiotic genes - their assemblage and localization in <u>Streptomyces</u>, in: "Overproduction of microbial metabolites", V. Krumphanzl, B. Sikyta, and Z. Vaněk, eds., Academic Press, London.

Sermonti, G., Petris, A., Micheli, M., and Lanfaloni, L., 1978, Chloramphenicol resistance in <u>Streptomyces coelicolor</u> A3(2) - possible involvement of a transposable element, <u>Mol.Gen.Genet.</u> 164:99.

Shaw, P.D., and Piwowarski, J., 1977, Effects of ethidium bromide and acriflavine on streptomycin-production by <u>Streptomyces bikiniensis</u>, <u>J.Antibiotics</u> 30:404.

Suter, M., Hütter, R., and Leisinger, T., 1978, Mutants of <u>Streptomyces glaucescens</u> affected in the production of extracellular enzymes, in: "Genetics of Actinomycetales, Proc.Internat.Coll. Borstel 1976", E. Freerksen, I. Tarnok, and J.H. Thumin, eds., Gustav Fischer Verlag, Stuttgart.

Toyama, H., Hayashi, E., Nojiri, Ch., Katsumata, K., Miyata, A., and Yamada, Y., 1982, Isolation and characterization of small plasmids from <u>Streptomyces</u>, <u>J.Antibiotics</u> 35:369.

Vivian, A., 1971, Genetic control of fertility in <u>Streptomyces coelicolor</u> A3(2) plasmid involvement in the interconversion of UF strains, <u>J.Gen.Microbiol.</u> 69:353.

Wright, L.F., and Hopwood, D.A., 1976, Actinorhodin is a chromosomally-determined antibiotic in <u>Streptomyces coelicolor</u> A3(2), <u>J.Gen.Microbiol.</u> 96:289.

NOVEL TECHNIQUES FOR BREEDING OF MICROORGANISMS AND APPLICATION OF THE MICROORGANISMS TO PRODUCTION OF GLUTATHIONE

Akira Kimura

Research Institute for Food Science
Kyoto University, Uji
Kyoto 611, Japan

Recent techniques such as gene engineering and/or cell fusion have made it possible to manipulate genes and to intentionally make microorganisms having various properties which we want. Using these useful techniques, we have succeeded in breeding powerful microorganisms which can produce high levels of ATP and/or glutathione (a detoxicating drug of liver). We also have developed novel techniques for the manipulation of the nucleus and the cytosol of yeasts. This article deals with these techniques and production of glutathione by the engineered strain of E. coli.

(I) TRANSFORMATION OF INTACT YEAST CELLS WITHOUT MAKING PROTOPLASTS (KU-METHOD)

Yeast cells could be transformed with plasmid DNA only when their protoplasts were employed[1]. However, preparation of the protoplasts and their subsequent regeneration of cell walls in solid medium was tedious and time consuming. Furthermore protoplasts often showed low regeneration efficiency. Therefore, development of a more convenient method for yeast tranformation has been sought. In our study on phosphorylation of various nucleotides by yeast cells, we could show that treatment of yeast cells with Triton X-100[2], a nonionic detergent, allowed them to incorporate various extracellular substances under conditions where Triton X-100 did not inhibit their growth[3]. This observation suggested that yeast cells, when treated with Triton X-100 or other detergents, might take up plasmid DNA. Therefore, we studied the uptake of plasmid DNA by intact yeast cells treated with various agents. Several alkali metal ions such as Li^+, Na^+, K^+, Cs^+, and Rb^+ were effective for inducing competence in yeast cells. Using intact yeast cells treated with these monovalent cations[4], we could attain a high transformation efficiencies which in some cases were comparable with that of the current protoplast method. Further, we found that 2-mercaptoethanol[5] and other thiol compounds such as LiSCN were also effective for yeast transformation[6]. This new method was named the "KU-method".

(1) Determination of the optimum conditions

(a) Effects of cations

The effects of various cations on transformation of S. cerevisiae D13-1A with plasmid YRp7 were studied. Among the cations tested, Li^+, Na^+, K^+, Cs^+, and Rb^+ were effective. Li-acetate was the most effective, as shown in Table I.

Table I. Effect of Various Metal Ions on Transformation.

Metal (0.1 M)	No of viable cells (/ml)	No of trans-formants (/10 μg DNA)
None	2.6×10^8	0
NaCl	2.3×10^8	540
KCl	2.1×10^8	280
RbCl	1.8×10^8	440
CsCl	2.3×10^8	590
$MgCl_2$	2.5×10^8	0
$CaCl_2$	2.0×10^8	0
$SrCl_2$	2.6×10^8	0
$BaCl_2$	2.4×10^8	0
$CoCl_2$	2.1×10^8	0
$MnCl_2$	2.6×10^8	0
$ZnSO_4$	1.3×10^8	0
$CuSO_4$	5.6×10^3	0
$Al_2(SO_4)_3$	1.9×10^8	0
None	1.6×10^8	0
LiCl	1.7×10^8	2330
Li-citrate	1.3×10^8	2290
Li-acetate	1.6×10^8	4540
Li_3PO_4 x	0.8×10^8	0
$LiNO_3$	1.6×10^8	2320
Li_2CO_3 x	1.0×10^6	0
Li_2SO_4	1.8×10^8	2410

x Not solubilized completely

(b) Concentration of metal ion and cells
The optimum concentration of Li$^+$ was found to be 0.1M, and that of cells to be 5 x 10^7 cells per ml, regardless of the plasmid DNA concentration. Transformation efficiency decreased above or below this cell concentration.
(c) pH and time of treatment
The optimum pH was near 7.0 and efficiency decreased above 8.0. The best duration of treatment was 1 hour, longer treatment causing death of the cells.
(d) Effects of temperature, polyethyleneglycol(PEG), and heat pulse
When cation treatment and transformation were carried out at 0 or 30 C, transformation efficiency was not greatly affected. However, the absence of PEG markedly decreased transformation efficiency. A heat pulse of 42 C for 5 min increased the appearance of transformants when the procedure was carried out at 30 C, but not at 0 C. However since heat pulse was not very important for transformation, we omitted it the rapid method described below.
(e) Effect of cell age and plasmid DNA concentration
The highest transformation efficiency was attained by the use of mid-log-phase cells grown on YPD medium. Increase of DNA concentration proportionally increased the efficiency of transformation.
(f) Effect of LiSCN on Transformation[6]
Recently we found LiSCN to be more effective than Li-acetate for transformation of some strains of S. cerevisiae(MT371a-7B,and DKD-5Dk-H).

(2) Uptake of linearized DNA

In addition to our study on the uptake of the circular plasmid YRp7, we investigated the uptake of YRp7 linearized with restriction endonucleases (BamHI and SalI) having a single susceptible site on plasmid YRp7. The linearized plasmid YRp7 was taken up by the metal-treated yeast cells, although its efficiency was much less than that of circular YRp7. The cleared lysates prepared from the Trp$^+$ yeast tranformants showed two bands equivalent to the covalently closed circular and open circular forms of the plasmid YRp7 standard. This fact suggested to us that the linearized plasmid DNA must be ligated to circular DNA in host cells.

(3) Transformation of various hosts with various plasmids

Using the KU-method, we compared transformation efficiencies with various combinations of hosts and vector systems. Various kinds of plasmids were used, for example, YEp6 which carried the HIS3 yeast structural gene and replicator derived from the 2μm DNA, and pDB248, which harbored the LEU2 yeast

Table II. Uptake of Various Plasmid DNAs by Yeast Cells Treated with Alkali Cations.

Recipient strain	Plasmid DNA	No of transformants(/10 µg DNA)
S. cerevisiae D13-1A	YRp7	4000 – 5000
S. cerevisiae D308.3	YRp7	2000
S. cerevisiae 108-3C	YRp7	2200
S. cerevisiae OK220-2C	YRp7	150 – 200
S. cerevisiae D13-1A	YCp19	520
S. cerevisiae D13-1A	YEp6	50
S. cerevisiae D13-1A	pGE742	40
S. cerevisiae D13-1A	pGS1	600 – 700
S. cerevisiae AH-22	pDB248	1000 – 2000

structural gene and replicator derived from 2 µm DNA, Table II shows that the efficiency varies depending on the host – vector combination. Based on the optimal conditions investigated, we proposed a protocol for yeast transformation (KU-method), which is shown in Fig. 1.

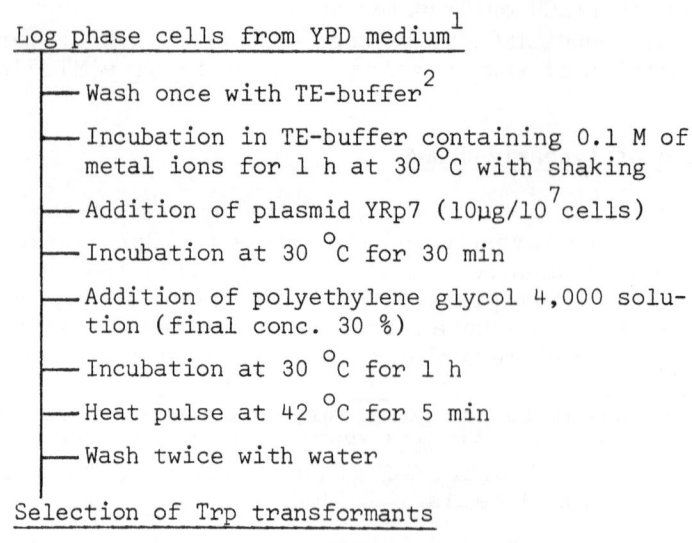

Log phase cells from YPD medium[1]

— Wash once with TE-buffer[2]

— Incubation in TE-buffer containing 0.1 M of metal ions for 1 h at 30 °C with shaking

— Addition of plasmid YRp7 (10µg/10[7]cells)

— Incubation at 30 °C for 30 min

— Addition of polyethylene glycol 4,000 solution (final conc. 30 %)

— Incubation at 30 °C for 1 h

— Heat pulse at 42 °C for 5 min

— Wash twice with water

Selection of Trp transformants

1) Saccharomyces cerevisiae D13-1A (a his3-532 trp1 gal1)
2) TE-buffer: 10 mM Tris-HCl buffer (pH 7.5) containing 1.0 mM EDTA

Fig. 1. Ku-Method for Transformation of Intact Yeast Cells Treated with Metal Ions.

(4) A rapid method for transformation

In order to simplify the KU-method, we added all the components at the same time, eliminating the heat shock process [6].The rapid process is very useful, although the yield decreases to almost one third. Figure 2 shows the rapid process, which is named the KUR-method[6].

Log phase cells from YPD medium

Wash once with TE-buffer

    ─Addition of plasmid DNA, alkali
     cation (final conc.50-100mM) and
     PEG4000(final conc.35%) all together

Incubation at 30 °C for 1-2hrs with shaking

Wash twice with water

Plate on selection medium and
Incubation at 30 °C for 2-3 days

Fig. 2. Kur-Method of Yeast Transformation with Alkali Cations.

(5) Mechanism of the introduction of plasmid DNA

Plasmid DNA was introduced only after PEG was added into the reaction mixture. Although PEG plays an important role in yeast transformation, its action mechanism is not clear yet.

Fig. 3.
Effect of Li$^+$ and Polyethylene
Glycol by Ellipticity of
    Plasmid DNA (YCp19).

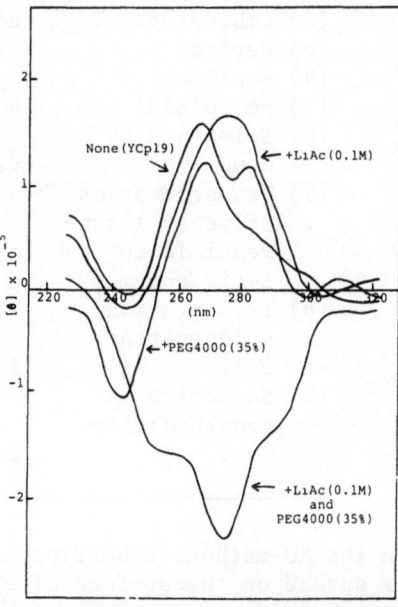

The highest number of transformants was obtained when the viable cell number began to decrease. Thus we felt that plasmid DNA entered the cell when its membrane became disordered just before cells were about to die. Since the CD-spectrum of plasmid DNA changed dramatically (Fig.3) when the DNA was mixed with PEG and Li-compounds, a configurational change of DNA might take place when DNA was introduced. Further study is now under way.

(6) Application of the KU-method

As an application of KU-method, we successfully isolated the glucokinase gene from yeast DNA[7]. Isolation of genes coding for enzymes of galactose metabolism by the use of this method is described by Fukazawa et al [8].

(7) Advantageous points of the KU-method

The KU-method has nine advantageous points, which are summarized in Table III.

1) In the KU-method, protoplasting is not necessary.

Table III. Advantageous Points of KU-Method.

	KU-method	Protoplast method
(1) Protoplast formation	unnecessary	necessary
(2) Embedding	unnecessary	necessary
(3) Period	2-3 days	5-7 days
(4) Replica	possible	impossible
(5) Polyploid	not appear	appear
(6) Selection by drug	very easy	difficult
(7) Transformation of yeast insusceptible to lytic enzymes	possible	impossible
(8) Preservation of competent cell	long time	short time
(9) Selection by growth difference	possible	impossible

2) In the KU-method, embedding is not necessary. Since transformants can be spread on the surface of agar plates, isolation or replication of them is very easy.

3) In the KU-method, transformants can be isolated in 2-3 days, while regeneration of cell walls took 5-7 days in the protoplast method.

4) Replication is possible in the KU-method directly from the selection agar plate. In the protoplast method it is impossible to replicate since protoplasts are embedded in the agar.

5) Since protoplasts fuse very easily, transformsants sometimes are polyploid, which is not desirable. However, in the KU-method no polyploid cells appear.

6) Since protoplasts are very fragile and sensitive to drugs, it is very difficult to select transformants by resistance to various drugs. In the KU-method, transformants can be selected by their resistance to drugs.

7) Protoplasts can be formed only from yeasts which are susceptible to lytic enzymes. Treatment of cells with Li-compounds in the KU-method is a physical process, applicable even to yeasts which are not susceptible to lytic enzymes.

8) The competent cells treated with Li-compounds can be preserved longer than protoplasts, which are fragile.

9) In the KU-method, transformants can be selected by the difference of colony size, since the colonies appear on the surface of agar. However, they cannot be selected on this basis in the protoplast method, since embedded colonies are smaller, larger or deformed, depending on how deep they are embedded.

(II) MINI-PROTOPLAST METHOD TO IMPROVE CYTOPLASM[9]

By the recombinant DNA technique, one can manipulate chromosomal DNA and change cellular properties. However, in some cases it is desirable to improve only cytoplasmic properties, leaving chromosomal properties as they were. We developed a new technique called the "mini-protoplast method", in which we could improve cytoplasmic properties by introducing only cytoplasmic elements of one cell into another. The mini-protoplasts can be prepared by treatment of budding yeast cells from the logarithmic phase with a lytic enzyme such as zymolyase. The mini-protoplasts are small protoplasts formed from buds, which contain at the beginning of budding only cytoplasmic materials such as mitochondria, vacuoles etc. Since they do not contain a nucleus, they cannot multiply. Using mini-protoplasts, we successfully transferred mitochondria of respiration sufficient cells to respiration-deficient ones (Table IV).

Since four transformants (No. 3-6) out of seven had the same nutrient requirements as the recipient and remained haploid, the transformants seemed to have the same cytoplasmic properties as the recipient. However, since they were respiration-sufficient, mitochondria must have been introduced into their cytoplasm.

Table IV. Characteristics of Mitochondrial Donor and Recipient Strains and Fusion products.

Strains	Mating-type	ρ	Auxotrophic markers					DNA content ($\mu g/10^8$ cells)	Ploidy
			ade2	arg4	leu2	his4	thr4		
(1) AN^{R}OR12D[a]	a	+	+	+	–	–	–	2.45	n
(2) BO60AF-1[b]	a	o	–	–	–	+	+	2.16	n
(3) F 1-1	a	+	–	–	–	+	+	2.24	n
(4) F 1-11	a	+	–	–	–	+	+	2.26	n
(5) F 2-12	a	+	–	–	–	+	+	2.55	n
(6) F 2-13	a	+	–	–	–	+	+	1.68	n
(7) F 2-2	a	+	+	+	–	+	+	4.44	$2n$
(8) F 2-4	a	+	+	+	–	+	+	4.50	$2n$
(9) F 2-11	a	+	+	+	–	+	+	4.10	$2n$

[a] mitochondrial donor
[b] mitochondrial recipient

Other transformants (No. 7-9) proved to be the usual fusants between two protoplasts, for they became diploid and their nutrient requirements were compensated.

This mini-protoplast method has proved useful for microbial breeding of yeasts. It was recently used for the breeding of Japanese sake yeasts with killer resistance. A good strain of S. cerevisiae (Kyoukai No. 7) whose properties were favorable for sake brewing, except for being killer-sensitive, was changed to killer-resistance leaving all the characteristic properties of sake making as they were. This method will be useful for future breeding of microorganisms.

(III) PRODUCTION OF GLUTATHIONE BY "SYNTECHNO SYSTEM"

Glutathione (GSH) is a detoxicating substance, and has many biochemical and pharmacological properties. It is a tripeptide, consisting of L-glutamate, L-cysteine, and glycine, and is biosynthesized by two steps as shown Fig. 4. The first step is catalyzed by an enzyme GSH-I, and the second by GSH-II. GSH-I is feedback inhibited by glutathione. Using E. coli C600, we tried to breed a strain of high potency. Our strategy is shown in Fig. 5. The first step was to obtain a gene coding for a desensitized GSH-I. We successfully isolated the gene and bound it to pBR322[11,12]. By the introduction of the hybrid plasmid pBR322-gsh-I to E. coli C600, we obtained colonies of high potency, which showed a red color on an agar plate with the nitroprusside-NH$_3$ reaction[11]. However, when the first step by GSH-I was released from negative feed-back inhibition and accelerated by the cloning of gsh-I, the second step catalyzed by

42

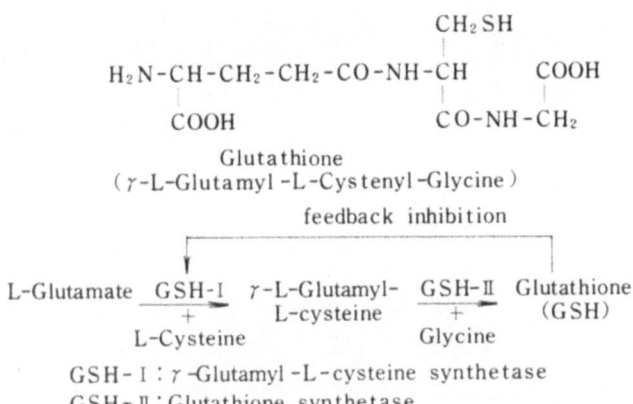

Fig. 4.   Structure and Biosynthetic Pathway of Glutathione.

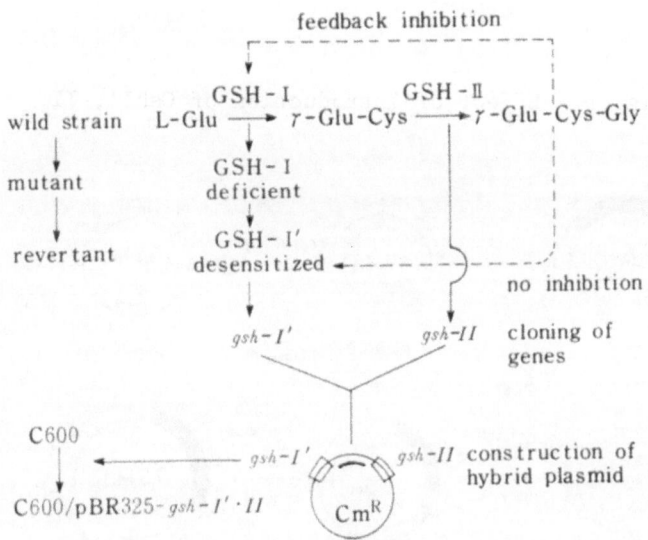

Fig. 5.   Strategy of Construction of an Engineered Cell.

GSH-II became rate-limiting: an intermediate γ-L-glutamyl-L-cysteine accumulated in the cells[12]. Therefore, we tried to clone the gene (gsh-II) coding for the second enzyme GSH-II, which was successfully obtained and bound to pBR322[13]. Although introduction of pBR322-gsh-II alone to original E. coli C600 was ineffective for glutathione production, its co-introduction together with gsh-I was quite effective. These results are shown in Fig. 6. From the engineered E. coli B, glutathione synthetase GSH-II was purified and characterized[14].

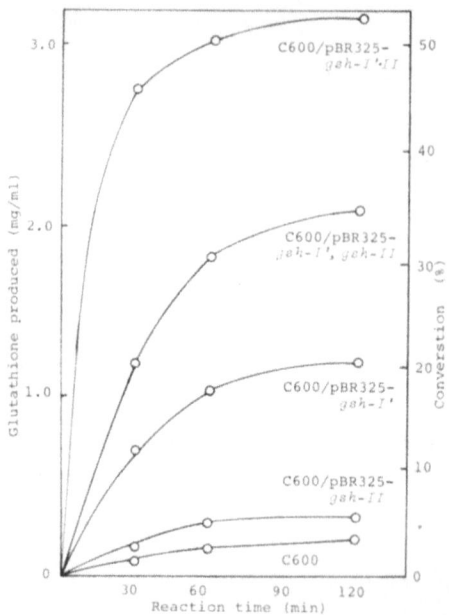

Fig. 6. Effect of Introduction of GshI', II.

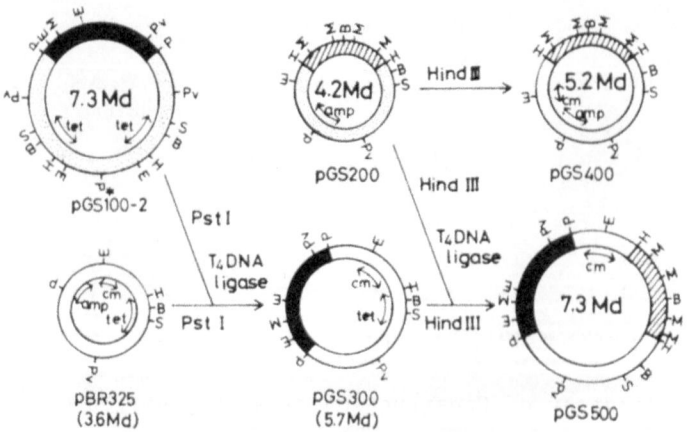

Scheme for construction of hybrid plasmids pGS300, pGS400, and pGS500. For the subcloning of the genes (*gsh I* for GSH-I, *gsh II* for GSH-II), two kinds of hybrid plasmids, pGS100-2 and pGS200, were used.   P* in pGS100-2 shows the site resistant to *Pst*I attack. The DNA fragment with *gsh I* was isolated from the *Pst*I digestion mixture of pGS100-2. pGS300 was constructed by inserting this DNA fragment into the *Pst*I site of pBR325. Similarly, the DNA fragment with the *gsh II* gene was isolated from the *Hind*III digestion mixture of pGS200. pGS400 and pGS500 were constructed by inserting this DNA fragment into the *Hind*III sites of pBR325 and pGS300, respectively. For other detailed conditions see Results. ■, *E. coli* B (RC912) chromosomal DNA fragment with *gsh I* gene; ▦, *E. coli* B (RC912) chromosomal DNA fragment with *gsh II* gene; ☐, vector plasmid pBR322; ☐, vector plasmid pBR325. Symbols: P, *Pst*I; E, *Eco*RI; B, *Bam*HI; S, *Sal*I; M, *Mlu*I; H, *Hind*III; Pv, *Pvu*II. *Amp, tet,* and *cm* show the genes for the resistance to ampicillin, tetracycline, and chloramphenicol, respectively.

Fig. 7.

Table V. Comparison of Engineered Cells.

Strain	Glutathione content in cells[1]	Enzyme activity[2]	
		GSH-I'	GSH-II
C600	0.84 (1.0)[3]	0.11(1.0)	0.24(1.0)
C600/pBR322-*gsh-I'*	1.24 (1.5)	1.31(11.9)	0.25(1.0)
C600/pBR322-*gsh-II*	0.98 (1.2)	0.12(1.1)	3.54(14.8)
C600/pBR325-*gsh-I'*	1.30 (1.5)	1.23(11.2)	0.24(1.0)
C600/pBR325-*gsh-II*	0.96 (1.1)	0.12(1.1)	3.26(13.5)
C600/[pBR325-*gsh-I'* / pBR325-*gsh-II*]	1.80 (2.1)	1.38(12.5)	0.91(3.8)
C600/pBR325-*gsh-I'·II*	1.89 (2.3)	1.50(13.6)	3.85(16.0)

1) mg/g wet wt. cells; 2) μmole/h/mg-protein
3) Relative content or activity

We next tried to make a hybrid plasmid containing both genes
gsh-I and gsh-II as shown in Fig.7 [15]. The engineered cells
which contained this hybrid plasmid produced a large amout of
glutathione from the individual component amino acids. Table V
and Fig. 6 show a comparison of the enzyme activities of GSH-I
and GSH-II, and the amount of glutathione produced by various
strains harboring gsh-I and/or gsh-II. Introduction of gsh-I and
gsh-II in one plasmid as pBR325-gsh-I·II was very effective,
while that of individual genes as pBR322-gsh-I and pBR322-gsh-II
was not. Besides the large amount of glutathione biosynthesized,

Table VI. Contribution of Various Techniques of GSH Production.

Contribution of Various Techniques to GSH Production

Strain	GSH (g/l)	Conversion from L-Cys	Notes	
C600	0.2	10%		
C600-gsh-I'	1.2	20		(1)
C600-gsh-II	0.3	10		
C600-gsh-I',II	2.2	35		(2)
C600-gsh-I'·II	3.3	52		(3)
"	4.5	75	L-Glu 80mM	
"	6.0	100	Imm. Cys 20mM	
"	11.3	94	" " 40	(4)
"	20.3	85	" " 80	
("	0.6	10	")	

(1) Fundamental knowledge on microorganisms.
(2) Gene engineering techniques.
(3) Fermentation techniques.
(4) Immobilization techniques.

the conversion rate (%) of L-cysteine to glutathione is very important, since L-Cys is the most expensive of the three component amino acids.

The conversion rate increased and reached even 100% when the engineered cells were immobilized with carrageenan gel. Looking at the Table VI, one might think that the gene engineering technique is not necessary, because the yield of glutathione increased from 0.2 to 3.2g/L by the gene engineering technique, yet was elevated to 25 g/L by other techniques such as fermentation and/or immobilization. However, when we tried to carry out the same experiment using original prototroph under the same conditions, we did not get such a high yield, i.e. 0.6 g/L, only 3 times as large as the control. Therefore, the combination of all these techniques (recombinant DNA, immobilization, fermentation) and fundamental knowledge (about negative feed-back inhibition) was very important. We propose to call such a combination "Syntechno system".

Recently we have succeeded in obtaining a hybrid plasmid, which contained two gsh-I and one gsh-II (symbolized as pBR325-gsh-I I II)[16]. Cells containing this plasmid biosynthesized glutathione much more effectively than the engineered cells mentioned above. The detailed data will be published elsewhere.

REFERENCE

(1) A. Hinnen, J. B. Hicks, and G. R. Fink, Transformation of Yeast, Proc. Natl. Acad. Sci. U.S.A., 75:1929 (1978)
(2) A. Kimura and M. Morita, Fermentative Formation of CDP-Choline by Intact Cells of a Yeast, Saccharomyces carlsbergensis (IFO 0641) Treated with a Detergent, Triton X-100, Agric. Biol. Chem., 39:1469 (1975)
(3) A. Kimura and K. Arima, and K. Murata, Biofunctional Change in Yeast Cell Surface on Treatment with Triton-X100, Agric. Biol. Chem., 45:2627 (1981)
(4) H. Ito, Y. Fukuda, K. Murata, and A. Kimura, Transformation of Intact Yeast Cells Treated with Alkali Cations, J. Bacteriol., 153:163 (1983)
(5) H. Ito, K. Murata, and A. Kimura, Transformation of Yeast Cells Treated with 2-mercaproethanol, Agric. Biol. Chem., 47:1691 (1983)
(6) H. Hashimoto, and A. Kimura, (unpublished data)
(7) Y. Fukuda, S. Yamaguchi, H. Hashimoto, and K. Kimura, (unpublished data)
(8) Y. Nogi, H. Hashimoto, and T. Fukazawa, Microbial. Gen. Genet. (in press) (1984)
(9) H. Fukuda and A. Kimura, Transfer of Mitochondria into Protoplasts of Saccharomyces cerevisiae by Mini-protoplast Fusion., FEBS Lett., 113:58 (1980)

(10) K. Murata and A. Kimura, Some Properties of Glutathione Biosynthesis-deficient Mutants of Escherichia coli B, J. Gen. Microbiol., 128:1047 (1982)

(11) K. Murata and A. Kimura, Cloning of a Gene Responsible for the Biosynthesis of Glutathione in Escherichia coli. Appl. Environ. Microbiol., 44:1444 (1982)

(12) H. Gushima, T. Miya, K. Murata, and A. Kimura, γ-Glutamylcysteine Production by Escherichia coli B dosed with the γ-Glutamylcysteine Synthetase Gene, Agric. Biol. Chem., 47:1927 (1983)

(13) K. Murata, T. Miya, H. Gushima, and A. Kimura, Cloning and Amplification of a Gene for Glutathione Synthetase in Escherichia coliB, Agric. Biol. Chem., 47:1381 (1983)

(14) H. Gushima, T. Miya, K. Murata, and A. Kimura, Purification and Characterization of Glutathione Synthetase from Escherichia coli B, J. Appl. Biochem., 5:210 (1983)

(15) H. Gusima, T. Miya, K. Murata, and A. Kimura, Construction of Glutathione-producing Strains of Escherichia coli B by Recombinant DNA Techniques, J. Appl. Biochem., 5:43 (1983)

(16) T. Miya, and A. Kimura (unpublished data)

AMINO ACID OVERPRODUCTION

Ralf Hütter and Peter Niederberger

Mikrobiologisches Institut
Eidgenössische Technische Hochschule
CH-8092 Zürich, Schweiz

INTRODUCTION

Many excellent articles have been written in the last years summarizing the "state of the art" in commercial production of amino acids (Nakayama et al., 1976; Hirose et al., 1978; Hirose and Shibai, 1980; Shiio, 1982, 1983; Nakayama, 1982; Kisumi, 1983; Tosaka et al., 1983). In nearly all cases bacteria are used for amino acid overproduction. It has become evident, that overproduction of any desired amino acid can be achieved by "breaking" the regulatory circuits, mainly by (a) eliminating feedback control acting on the activity of entry enzymes and on enzyme levels, (b) enhancing the availability of precursors by e.g. eliminating or reducing reactions · competing for common intermediates, and (c) eliminating degradation. Which specific way of selecting overproducers is chosen, is of course highly dependent on the organism and amino acid of interest. In general organisms with a relatively simple regulatory set-up, with no or few isozymes and with concerted feedback inhibition of entry enzymes are preferred (see e.g. Nakayama et al., 1976; Hütter, 1979; Tosaka et al., 1983). Only recently has gene technology been applied to amino acid overproduction in bacteria, primarily with Escherichia coli for threonine biosynthesis (Debabov, 1983; Miwa et al., 1981; Momose, 1983) and tryptophan biosynthesis (Tribe and Pittard, 1979; Aiba et al., 1980, 1983; Tribe et al., 1982). In all these cases the corresponding biosynthetic operons were cloned onto multicopy plasmids, thus achieving high enzyme levels, i.e. high biosynthetic capacities.

49

This report will discuss problems with amino acid production in eukaryotic microorganisms and then focus on tryptophan biosynthesis in the yeast <u>Saccharomyces</u> <u>cerevisiae</u>.

AMINO ACID BIOSYNTHESIS IN EUKARYOTIC MICROORGANISMS, MAINLY FUNGI

In several respects regulation of amino acid biosynthesis is similar in pro- and eukaryotes (feedback inhibition of activity of entry enzymes by end products; availability of precursors; degradation), but it differs in others (genetic organization; control of enzyme levels; promoter regions; compartmentation). These aspects were recently discussed by Hütter et al. (1982a,b; 1983a), and will be further dealt with below. Compartmentation, however, will not be treated in this review; the interested reader is referred to Davis (1975).

The most outstanding features of amino acid biosynthesis in fungi, in comparison to bacterial systems and mainly enterobacteria, are the scattering of the structural genes of a given biosynthetic sequence (e.g. for <u>S.cerevisiae</u>, see Mortimer and Schild, 1980), the lack of pathway specific controls (except for single enzymes) and the presence of a system of "general control of amino acid biosynthesis" or "metabolic interlock". This regulatory system has been described for the yeasts <u>Saccharomyces</u> <u>cerevisiae</u>, <u>Hansenula henricii</u> and a <u>Candida</u> species (Schürch et al., 1974; Greer and Fink, 1975; Delforge et al., 1975; Wolfner et al., 1975; Wiame and Dubois, 1976; Miozzari et al., 1978; Bode and Birnbaum, 1981; Bode et al., 1983), for <u>Neurospora</u> <u>crassa</u> (Carsiotis and Jones, 1974; Carsiotis et al., 1974) and for <u>Aspergillus</u> <u>nidulans</u> (Piotrowska, 1980). The "general control" is characterized by parallel derepression of enzymes of diverse biosynthetic pathways in response to various amino acid limitations; derepression factors of 2 to 4 are usually observed. As a consequence of the scattering of the structural genes and differences in promoter structures (Hinnebusch and Fink, 1983; Dobson et al., 1983), regulation of synthesis of enzymes of a given pathway is usually not or only moderately coordinate (Jones and Fink, 1982). The system enables the cells to cope efficiently with amino acid imbalances in the growth medium, a situation likely occurring in the natural habitat of fungi such as yeasts (Niederberger et al., 1981, 1983).

Fungi maintain considerably higher levels of most amino acid biosynthetic enzymes than needed for maximal growth speed. No influence on growth rate is found upon reduction of single enzyme levels to 1/2 or even to 1/4 of the wild type level (Hilger et al., 1973; Miozzari et al., 1978). This means that most enzymes have low "sensitivity coefficients", and flux is largely buffered against

downward modulations (Flint et al., 1981). Reduction under a critical level, however, leads to a very low flux distal to the limiting
step. Thus only under artificial deregulation, where the maximal
flux capacity is exploited, a certain enzyme may become rate limiting in practical terms.

TRYPTOPHAN BIOSYNTHESIS IN THE YEAST SACCHAROMYCES CEREVISIAE

        Tryptophan biosynthesis occurs through the same biosynthetic
pathway as in other microorganims (see Gibson and Pittard, 1968;
Hütter, 1973; Crawford, 1975, 1980; Crawford and Stauffer, 1980).
Five genes, scattered over four chromosomes, encode the tryptophan
biosynthetic enzymes (Table 1).

Table 1. Tryptophan Biosynthetic Pathway, Enzymes and Genes Involved
          in Saccharomyces cerevisiae

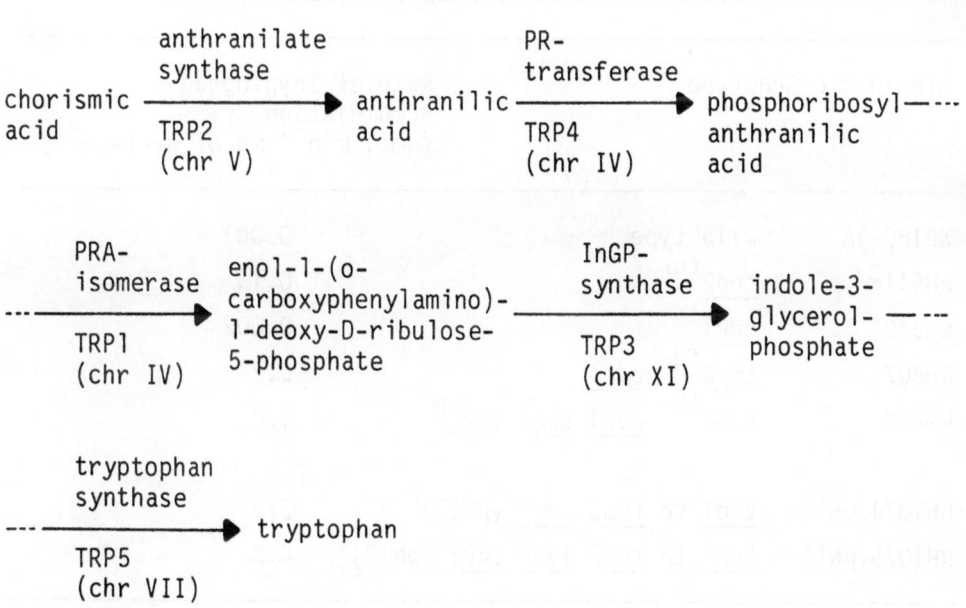

        The flux through the pathway is primarily regulated through
feedback inhibition of the entry enzyme anthranilate synthase and by
the availability of the precursor chorismic acid. No repression of
enzyme levels by the end product is observed, and artificial downward modulation of individual enzyme levels influences the flux only
slightly. Derepression of enzyme levels, on the other hand, is only
moderate and only four of the five tryptophan biosynthetic enzymes

are derepressible under the action of the "general control" system (Kradolfer et al., 1977, 1982; Miozzari et al., 1978; Hütter et al., 1982a).

Tryptophan accumulation can be considerably enhanced by the introduction of a set of mutations into the wild type strain, namely (a) feedback insensitive anthranilate synthase ($trp2^{fbr}$), (b) constitutively derepressed enzyme levels (cdr1 or cdr2), (c) defective chorismate mutase (aro7) eliminating competition for the precursor chorismic acid, and (d) defective aromatic aminotransferase (aat2) reducing tryptophan degradation (Table 2, first 5 entries; Hütter et al., 1983a; Kradolfer et al., 1982).

Table 2. Rate of Tryptophan Accumulation

Strain and Genotype		Rate of Tryptophan Accumulation (nmol min^{-1} mg of protein^{-1})
X2180-1A	wild type	0.001
RH511	$trp2^{fbr}$	0.13
RH558	$cdr1^{a}$	0.002
RH807	$trp2^{fbr}$ cdr1	0.34
RH813	$trp2^{fbr}$ cdr1 aro7 aat2^{b}	1.0
RH1074.pNI7	trp1 to trp5 leu2 (pNI7)^{c}	3.2
RH1075.pNI7	trp1 to trp5 leu2 cdr2 (pNI7)^{c}	4.8

[a]Strains carrying mutation cdr2-1 show comparably low accumulation rates.

[b]Mutation aro7 leads to a defective chorismate mutase (Kradolfer et al., 1977); mutation aat2 leads to a defective aromatic aminotransferase (Kradolfer et al., 1982).

[c]Trp accumulation rates after addition of anthranilic acid (10^{-4}M).

The limits for further increases in flux and consequently the rate of tryptophan accumulation are enzyme levels. Several possibilities can be envisaged to overcome this problem. Higher enzyme levels could be obtained by (a) increasing gene dosage, (b) selecting for up-promoters, or (c) by linking genes to more efficient promoters. We have used the first approach in which all five genes were cloned on a multicopy vector (for details see below), and which allows for further increase in tryptophan accumulation (Table 2, last two entries).

Table 3. Relative Enzyme Levels in Yeast Strains Carrying Hybrid Plasmids With Tryptophan Biosynthetic Genes

Strain and Genotype[a]		Relative Enzyme Levels[b]		
		Anthranilate synthase	PRA-isomerase	InGP-synthase
X2180-1A	wild type	1	1	1
RH974.pNI1[c]	trp3 leu2 (pNI1)	1.5	1.5	110
RH1074.pNI7	trp1 to trp5 leu2 (pNI7)	7.9	1.6	7.6
RH1075.pNI7	trp1 to trp5 leu2 cdr2 (pNI7)	39.6	2.0	29.3

[a] trp and leu stand for tryptophan and leucine auxotrophy. cdr = constitutively derepressed enzyme levels.

[b] absolute levels are for anthranilate synthase 1.05, for PRA-isomerase 4.1 and for indole glycerol phosphate synthase 0.94 nmol min^{-1} mg protein^{-1}.

[c] pNI1 corresponds to a plasmid carrying the gene TRP3. For pNI7 see Fig. 1.

Two of the five tryptophan biosynthetic genes have previously been cloned by other groups, TRP5 by Walz et al. (1978) and TRP1 by Struhl et al. (1979). We have cloned the remaining three genes TRP2 and TRP3 (Aebi et al., 1982) and TRP4 (unpublished data). When carried on a multicopy plasmid the enzyme levels correspond quite well to gene dosage, i.e. copy number of the hybrid plasmid, and the genes are still subject to the "general control" (Aebi et al.,

1982; Hütter et al., 1983b), with the exception of the TRP1 gene (see also Table 3).

Based on the singly cloned genes a multicopy plasmid pNI7 was constructed carrying all five tryptophan biosynthetic structural genes (Fig. 1; Hütter et al., 1983b, 1983c). The plasmid pNI7 was introduced into a yeast strain carrying chromosomal mutations for trp1 to trp5 and leu2, in order to stabilize plasmid maintenance. Such a strain (RH1075.pNI7) shows a considerable enhancement of tryptophan accumulation. In the experiments shown (Table 3) the strain carrying pNI7 had to be fed for the accumulation studies with anthranilate, because pNI7 contains a wild type TRP2 gene, coding for a feedback sensitive anthranilate synthase. The accumulation rate is furthermore limited by the fact, that PRA-isomerase (TRP1 gene product) seems to be expressed very poorly on the plasmid and furthermore does not respond to the "general control" elicited by the cdr2 mutation.

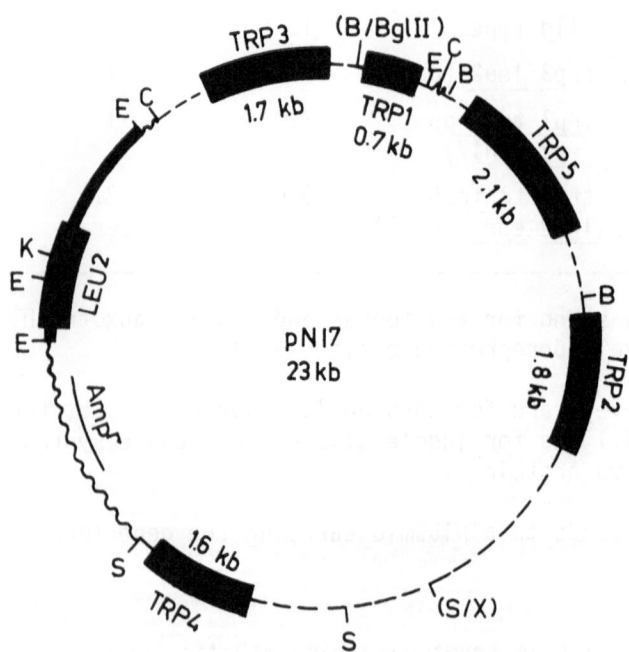

Fig. 1. Plasmid pNI7 Carrying the Five Tryptophan Biosynthetic Genes of Yeast

▬▬▬ cloned yeast DNA fragments carrying tryptophan and LEU2 structural genes, – – – additional yeast DNA, ▬▬ yeast 2 μm DNA, ∿∿∿ pBR322 sequences.

For further improvement it is crucial to introduce a mutant trp2 gene, coding for a feedback insensitive anthranilate synthase, to overcome the poor expression of PRA-isomerase and then to guarantee sufficient availability of the precursor chorismic acid.

ACKNOWLEDGEMENT

The work with S.cerevisiae was supported by research grants from the Swiss National Science Foundation (Nr. 3.262.078 and Nr. 3.002.081).

REFERENCES

Aebi, M., Niederberger, P., and Hütter, R., 1982, Isolation of the TRP2 and the TRP3 genes of Saccharomyces cerevisiae by functional complementation in yeast, Curr.Genet. 5:39.

Aiba, S., Imanaka, T., and Tsunekawa, H., 1980, Enhancement of tryptophan production by Escherichia coli as an application of genetic engineering, Biotech.Lett. 2:525.

Aiba, S., Tsunekawa, H., and Imanaka, T., 1983, Genetic manipulation for tryptophan production by Escherichia coli, in: "Proc. of the IVth Internat.Symp. on Genetics of Industrial Microorganisms (GIM 82)", Y. Ikeda, and T. Beppu, eds., Kodansha Ltd., Tokyo.

Bode, R., and Birnbaum, D., 1981, Allgemeine Regulation der Aminosäurebiosynthese in Hansenula henricii, Z.Allg.Mikrobiol. 21: 719.

Bode, R., Casper, P., and Kunze, G., 1983, Induction of the general control of amino acid biosynthesis in Candida spec. EH 15/D using amitrole, Biochem.Physiol.Pflanzen 178:457.

Carsiotis, M., and Jones, R.F., 1974, Cross-pathway regulation: tryptophan-mediated control of histidine and arginine biosynthetic enzymes in Neurospora crassa, J.Bacteriol. 119:889.

Carsiotis, M., Jones, R.F., and Wesseling, A.C., 1974, Cross-pathway regulation: histidine-mediated control of histidine, tryptophan, and arginine biosynthetic enzymes in Neurospora crassa, J.Bacteriol. 119:893.

Crawford, I.P., 1975, Gene rearrangements in the evolution of the tryptophan synthetic pathway, Bacteriol.Rev. 39:87.

Crawford, I.P., 1980, Comparative studies on the regulation of tryptophan synthesis, CRC Crit.Rev.Biochem. 8:175.

Crawford, I.P., and Stauffer, G.V., 1980, Regulation of tryptophan biosynthesis, Ann.Rev.Biochem. 49:163.

Davis, R.H., 1975, Compartmentation and regulation of fungal metabolism: Genetic approaches, Ann.Rev.Genet. 9:39.

Delforge, J., Messenguy, F., and Wiame, J.M., 1975, Regulation of arginine biosynthesis in Saccharomyces cerevisiae. The specificity of argR⁻ mutations and the general control of amino acid biosynthesis, Eur.J.Biochem. 57:231.

Debabov, V., 1983, Construction of strains producing L-threonine, in: "Proc. of the IVth Internat.Symp. on Genetics of Industrial Microorganisms (GIM 82)", Y. Ikeda, and T. Beppu, eds., Kodansha Ltd., Tokyo.

Dobson, M.J., Tuite, M.F., Mellor, J., Roberts, N.A., King, R.M., Burke, D.C., Kingsman, A.J., and Kingsman, S.M., 1983, Expression in Saccharomyces cerevisiae of human interferon-alpha directed by the TRP1 5'-region, Nucl.Acid.Res. 11:2287.

Flint, H.J., Tateson, R.W., Barthelmess, I.B., Porteous, D.J., Donachie, W.D., and Kacser, H., 1981, Control of the flux in the arginine pathway of Neurospora crassa, Biochem.J. 200:231.

Gibson, F., and Pittard, J., 1968, Pathways of biosynthesis of aromatic amino acids and vitamins and their control in microorganisms, Bacteriol.Rev. 32:465.

Greer, H., and Fink, G., 1975, Isolation of regulatory mutants in Saccharomyces cerevisiae, in: "Methods in Cell Biology, Vol. XI", D.M. Prescott, ed., Academic Press, New York.

Hilger, F., Culot, M., Minet, M., Piérard, A., Grenson, M., and Wiame, J.-M., 1973, Studies on the kinetics of the enzyme sequence mediating arginine synthesis in Saccharomyces cerevisiae, J.Gen.Microbiol. 75:33.

Hinnebusch, A.G., and Fink, G.R., 1983, Repeated DNA-sequences upstream from HIS1 also occur at several other co-regulated genes in Saccharomyces cerevisiae, J.Biol.Chem. 258:5238.

Hirose, Y., Sano, K., and Shibai, H., 1978, Amino acids, Ann.Rept. Ferm.Proc. 2:155.

Hirose, Y., and Shibai, H., 1980, Amino acid fermentation, Biotech. Bioeng. 22:111.

Hütter, R., 1973, Regulation of the tryptophan biosynthetic enzymes in fungi, in: "Proc.1st Internat.Symp. on Genetics of Industr. Microorganisms (GIM 70)", Z. Vaněk, Z. Hošťálek, and J. Cudlín, eds., Academia, Prague.

Hütter, R., 1979, Regulation of primary metabolism, in: "Proc. 3d Internat.Symp. on Genetics of Industr. Microorganisms (GIM 78), O.K. Sebek and A.I. Laskin, eds., American Society for Microbiology, Washington.

Hütter, R., Aebi, M., Kradolfer, P., and Niederberger, P., 1982a, Specific and general control systems in microbial amino acid biosynthesis, in: "Proc.FEMS Symposium on Overproduction of Microbial Metabolites", V. Krumphanzl, B. Sikyta, and Z. Vaněk, eds., Academic Press, London.

Hütter, R., Niederberger, P., and Kradolfer, P., 1982b, Regulation of primary metabolism, in: "Proc.6th Internat.Ferment. Symposium "Advances in Biotechnology", Vol. 3", M. Moo-Young, C. Vezina, and K. Singh, eds., Pergamon Press, Toronto.

Hütter, R., Niederberger, P., and Aebi, M., 1983a, Regulation of amino acid biosynthesis in eucaryotic microorganisms, in: "Proc.4th Internat.Symp. on Genetics of Industr.Microorganisms (GIM 82)", Y. Ikeda, and T. Beppu, eds., Kodansha Ltd., Tokyo.

Hütter, R., Niederberger, P., Aebi, M., Furter, R., and Prantl, F., 1983b, Cloning of tryptophan biosynthetic genes of yeast in yeast. Construction of composite plasmid carrying all five structural genes, Riv.Biol. 76:207.

Hütter, R., Niederberger, P., Aebi, M., Prantl, F., and Furter, R., 1983c, Cloning of yeast tryptophan biosynthetic genes in yeast construction of an artificial gene cluster and influence on enzyme levels and flux through the pathway, in: "Proc.3rd Symp.Soc.Countries on Biotechnology, Bratislava" (in press).

Jones, E.W., and Fink, G.R., 1982, Regulation of amino acid and nucleotide biosynthesis in yeast, in: "The molecular biology of the yeast Saccharomyces", J.N. Strathern, E.W. Jones, and J.R. Broach, eds., Cold Spring Harbor Laboratory, Cold Spring Harbor, New York.

Kisumi, K., 1983, Transductional construction of amino acid-producing strains of Serratia marcescens, in: "Proc.4th Internat. Symp. on Genetics of Industr. Microorganisms (GIM 82)", Y. Ikeda, and T. Beppu, eds., Kodansha Ltd., Tokyo.

Kradolfer, P., Zeyer, J., Miozzari, G., and Hütter, R., 1977, Dominant regulatory mutants in chorismate mutase of Saccharomyces cerevisiae, FEMS Microbiol.Lett. 2:211.

Kradolfer, P., Niederberger, P., and Hütter, R., 1982, Tryptophan degradation in Saccharomyces cerevisiae: Characterization of two aromatic aminotransferases, Arch.Microbiol. 133:242.

Miozzari, G., Niederberger, P., and Hütter, R., 1978, Tryptophan biosynthesis in Saccharomyces cerevisiae: Control of the flux through the pathway, J.Bacteriol. 134:48.

Miwa, K., Tsuchida, T., Kurahashi, O., Nakamori, S., Sano, K., and Momose, H., 1981, Genetic construction of L-threonine-producing strains by self-cloning of E.coli K-12 mutants, in: "Proc. Agric.Chem.Soc.Japan", ex Enei, H., Shibai, H., and Hirose, Y., 1982, Amino acids and nucleic acid-related compounds, Ann.Rept. Ferm.Proc. 5:79.

Momose, H., 1983, New genetic approaches to amino-acid-producing strains, Dev.Industr.Microbiol. 24:109.

Mortimer, R.K., and Schild, D., 1980, Genetic map of Saccharomyces cerevisiae, Microbiol.Rev. 44:519.

Nakayama, K., 1982, Breeding of amino acid producing micro-organisms, in: "Proc.FEMS Symposium on Overproduction of Microbial Metabolites", V. Krumphanzl, B. Sikyta, and Z. Vaněk, eds., Academic Press, London.

Nakayama, K., Araki, K., Hagino, H., Kase, H., and Yoshida, H., 1976, Amino acid fermentations using regulatory mutants of Corynebacterium glutamicum, in: "Proc.2nd Internat.Symp. on Industr. Microorganisms (GIM 74)", K.D. Macdonald, ed., Academic Press, New York.

Niederberger, P., Miozzari, G., and Hütter, R., 1981, Biological role of the general control of amino acid biosynthesis in Saccharomyces cerevisiae, Mol.Cell.Biol. 1:584.

Niederberger, P., Aebi, M., and Hütter, R., 1983, Influence of the general control of amino acid biosynthesis on cell growth and cell viability in Saccharomyces cerevisiae, J.Gen.Microbiol. 129:2571.

Piotrowska, M., 1980, Cross-pathway regulation of ornithine carbamoyltransferase synthesis in Aspergillus nidulans, J.Gen. Microbiol. 116:335.

Schürch, A., Miozzari, G., and Hütter, R., 1974, Regulation of tryptophan biosynthesis in Saccharomyces cerevisiae: mode of action of 5-methyltryptophan and 5-methyltryptophan sensitive mutants, J.Bacteriol. 117:1131.

Shiio, I., 1982, Metabolic regulation and over-production of amino acids, in: "Proc.FEMS Symposium on Overproduction of Microbial Metabolites", V. Krumphanzl, B. Sikyta, and Z. Vaněk, eds., Academic Press, London.

Shiio, I., 1983, Biochemical conditions for amino acid overproduction, in: "Proc.4th Internat.Symp. on Genetics of Industr. Microorganisms (GIM 82)", Y. Ikeda, and T. Beppu, eds., Kodansha Ltd., Tokyo.

Struhl, K., Stinchcomb, D.T., Scherer, S., and Davis, R.W., 1979, High-frequency transformation of yeast: autonomous replication of hybrid DNA molecules, Proc.Natl.Acad.Sci.USA 76:1035.

Tosaka, O., Enei, H., and Hirose, Y., 1983, The production of L-lysine by fermentation, TIB 1:70.

Tribe, E., and Pittard, J., 1979, Hyperproduction of tryptophan by Escherichia coli: genetic manipulation of the pathways leading to tryptophan formation, Appl.Environm.Microbiol. 38:181.

Tribe, D.E., Choi, Y.J., Cuevas, C.A., and Pittard, A.J., 1982, Aromatic amino acid hyperproducing mutants of Escherichia coli: biochemistry, genetics and fermentation kinetics, in: "Advances in Biotechnology, Proc.6th Internat.Ferment.Symp.", M. Moo-Young, C. Vezina, and K. Singh, eds., Pergamon Press, Toronto.

Walz, A., Ratzkin, R., and Carbon, J., 1978, Control of expression of a cloned yeast (Saccharomyces cerevisiae) gene (trp5) by a bacterial insertion element (IS2), Proc.Natl.Acad.Sci.USA 75: 6172.

Wiame, J.-M., and Dubois, E.L., 1976, The regulation of enzyme synthesis in arginine metabolism in Saccharomyces cerevisiae, in: "Proc.2nd Internat.Symp. on Genetics of Industr. Microorganisms (GIM 74)", K.D. Macdonald, ed., Academic Press, New York.

Wolfner, M., Yep, D., Messenguy, F., and Fink, G.R., 1975, Integration of amino acid biosynthesis into the cell cycle of Saccharomyces cerevisiae, J.Mol.Biol. 90:273.

THE DYNAMIC ANALYSIS OF METABOLITE

ACCUMULATION IN MICROORGANISMS

Barbara E. Wright

Department of Microbiology
University of Montana
Missoula, MT 59812

The metabolism underlying industrial fermentations, secondary metabolism, starvation metabolism, encystment, sporulation, and microbial differentiation have much in common. All of these metabolic transitions are to various degrees the consequence of a decreased rate of growth, due to a nutritional limitation or partial starvation. The residual kind of "unbalanced metabolism" results in the accumulation of some new material which was present but had not accumulated during cell growth. In order to influence the yield of the material which accumulates – or the rate at which it is formed, it is necessary to understand which reactions are most rate-limiting during and after the metabolic transition from growth to partial starvation. The logical approach to this problem is to use a dynamic systems analysis, in which metabolism in vivo is analyzed by computer simulation. This approach will be explained following a brief comparison of various metabolic transitions which are probably under the control of very similar biochemical mechanisms.

Figure 1 indicates the consequences of nitrogen-limitation on metabolite accumulation in Agrobacter.[1] As growth slows, uridine diphosphoglucose (UDPG) and glycogen accumulate; when a nitrogen source is added, growth resumes and these metabolites decrease in concentration. A similar situation is seen in yeast as the available glucose is used and cell division stops (Figure 2).[2] Trehalose accumulates, but then decreases in concentration when glucose is again added to the culture (arrow). Two industrial fermentations behave in an analogous fashion: citric acid accumulation following glucose exhaustion in Aspergillus,[3] and penicillin formation, which also results from the depletion of exogenous glucose (Figure 3).[4]

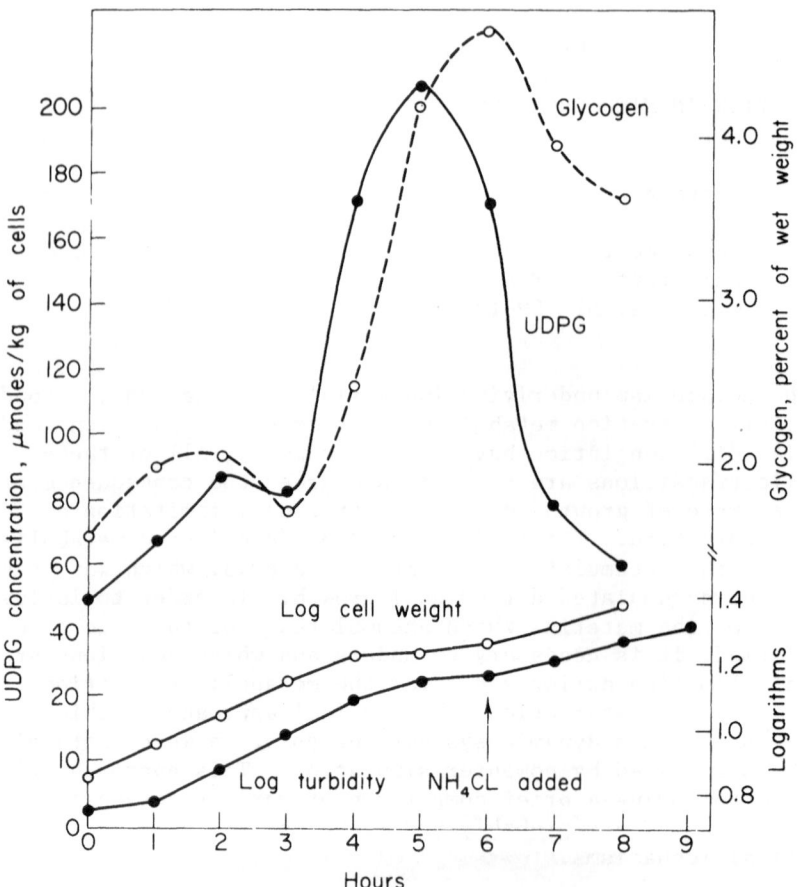

Fig. 1. The accumulation and disappearance of uridine diphosphoglucose (UDPG) and glycogen in <u>Agrobacter</u>, as a function
of nitrogen-limitation.

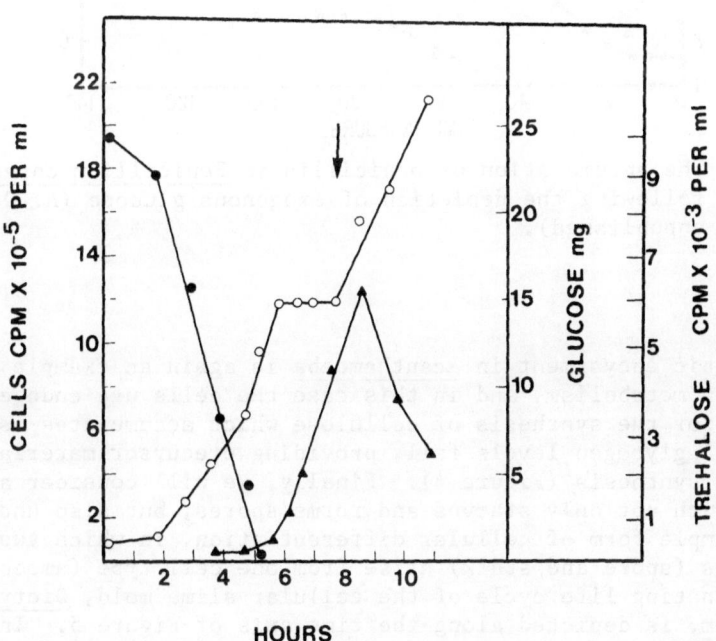

Fig. 2   The accumulation of trehalose (triangles) in yeast (open
circles), as a function of the availability of glucose
(closed circles).[2]

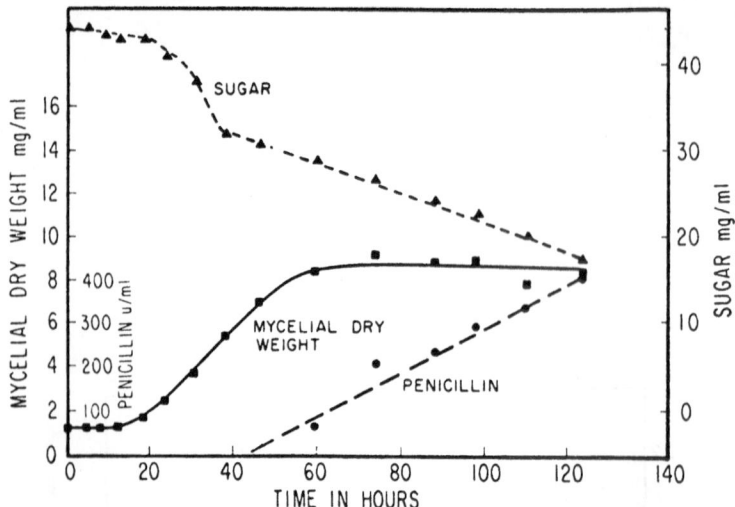

Fig. 3. The accumulation of penicillin in _Penicillium_ chrysogenum, following the depletion of exogenous glucose (A. Demain, unpublished).

Amoebic encystment in _Acanthamoeba_ is again an example of starvation metabolism, and in this case the cells use endogenous reserves for the synthesis of cellulose which accumulates as cysts are made;[5] glycogen levels fall, providing precursor material for cellulose synthesis (Figure 4). Finally, we will consider a system which not only starves and forms spores, but also undergoes a very simple form of cellular differentiation, in which two new cell types (spore and stalk) arise from one cell type (amoebae).[6] The fascinating life cycle of the cellular slime mold, _Dictyostelium discoideum_, is depicted along the time axis of Figure 5. In the presence of external nutrients the organism grows and multiplies as free-living single amoebae. Under starvation conditions, growth ceases and the amoebae aggregate to form a multicellular colony which differentiates through several distinctive stages to form a sorocarp or fruiting body. During differentiation there is no net loss of carbohydrate. At aggregation most of the carbohydrate material is present as glycogen and RNA (not shown). In the middle of differentiation there occurs a transient accumulation of UDPG, glucose-6-phosphate and other similar metabolites. At the end of differentiation there is a net decrease in these macromolecules and a comparable increase in new polysaccharide end-products (trehalose, cellulose-glycogen wall complex and mucopolysaccharide).[6] For the purposes of the present discussion, let us assume that trehalose accumulation is an important industrial process.

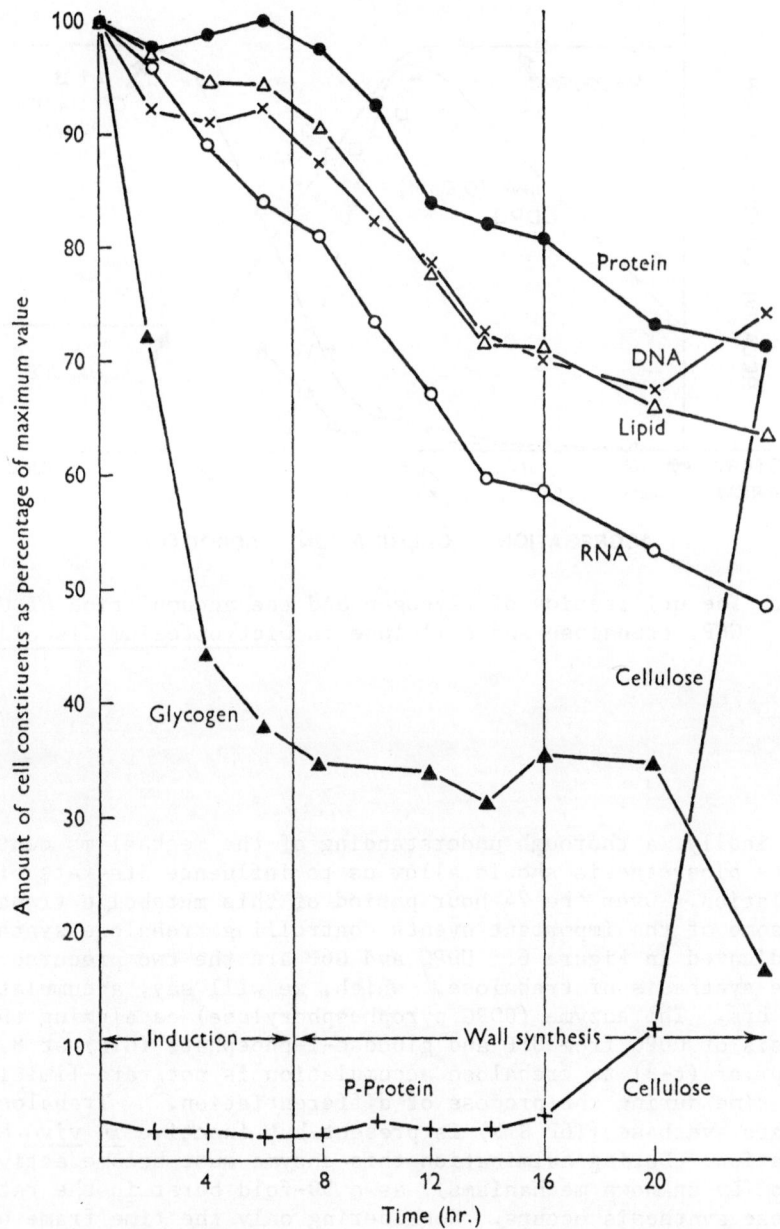

Fig. 4. The utilization of glycogen and the accumulation of cellulose in _Acanthamoeba_, in response to starvation during
encystment.

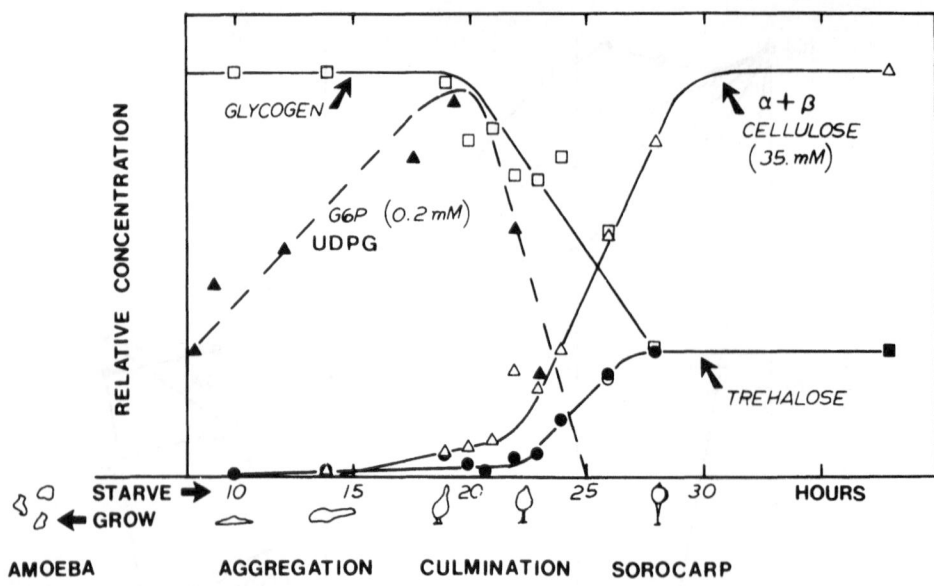

Fig. 5. The utilization of glycogen and the accumulation of UDPG,[6] G6P, trehalose and cellulose in <u>Dictyostelium discoideum</u>.

Theoretically, a thorough understanding of the mechanisms control-ling its biosynthesis should allow us to influence its rate of accumulation. Over the 24 hour period of this metabolic transi-tion, some of the important events controlling trehalose synthesis are indicated in Figure 6. UDPG and G6P are the two precursors for the synthesis of trehalose,[7] which, we will say, accumulates at t=0 hrs. The enzyme (UDPG pyrophosphorylase) catalyzing the synthesis of UDPG from UTP and glucose-1-phosphate (G1P) at 8 hours prior (t-8) to trehalose accumulation is not rate-limiting at any time during the process of differentiation.[8] Trehalose-6-phosphate synthase (T6P SYN) is present but inactive <u>in vivo</u> at aggregation. During culmination this enzyme must become activated <u>in vivo</u> (by unknown mechanisms), as a 50-fold burst in the rate of trehalose synthesis occurs. Considering only the time frame of about 10 hours prior (t-10) to trehalose accumulation, mechanisms controlling protein synthesis are not critical variables for end product synthesis; rather, metabolite (G1P, UTP, G6P, UDPG) flux and the mechanisms unmasking T6P synthase activity are rate-limiting in this system. In earlier or broader time frames, amino acid oxidation, ATP synthesis, net endogenous protein and glycogen

66

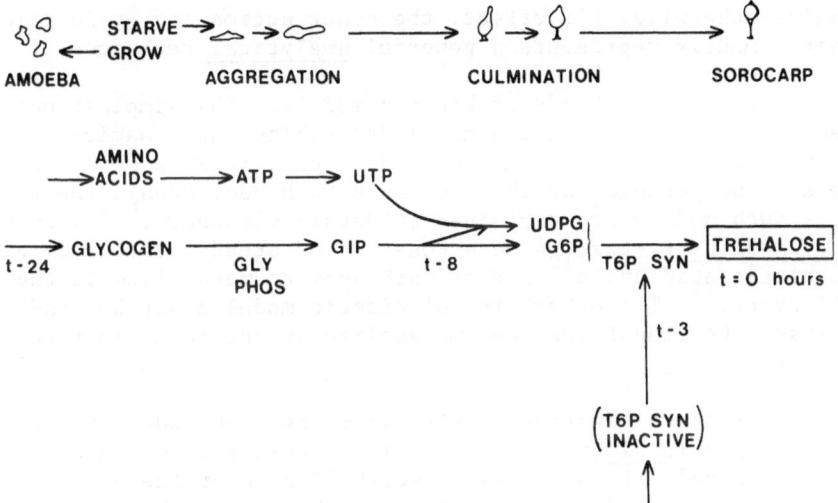

Fig. 6. Some of the areas of metabolism and reaction controlling
trehalose accumulation during differentiation in
<u>Dictyostelium</u> <u>discoideum</u>. See text for further details.

breakdown, UTP production, the synthesis and degradation of
glycogen phosphorylase, and so on, will also be critical variables. Thus, the selection of a time frame is an essential
prerequisite to describing those events controlling trehalose
accumulation.

In order to understand the biochemical mechanisms underlying
trehalose accumulation, it is necessary to obtain the data required
for the construction of kinetic models, which represent one of the
few noninvasive techniques available for analyzing metabolism <u>in</u>
<u>vivo</u>. As our biochemical knowledge grows, models will become
increasingly essential to its integration and comprehension. By
organizing the data, models serve to clarify which facts are com-
patible and pinpoint those which are not. Models help us ask ques-
tions and think in realistic, dynamic terms which relate directly
to the living organism. They serve as a realistic framework
within which to judge the relevance of <u>in vitro</u> data to metabolism
<u>in vivo</u>, make specific predictions to be tested experimentally,
and suggest analytical approaches that would be difficult to

imagine otherwise. In effect, the construction and exploration of kinetic models represents a powerful _analytical_ technique.

There are two kinds of kinetic models. The simplest one is a steady state model; i.e., a model describing the dynamics of metabolism at one point in time under conditions in which pool sizes do not change and the flux into each pool equals the flux out. Such models are explained in detail elsewhere.[8-11] In the _Dictyostelium_ system, we have constructed steady state models of the citric acid cycle[10] and of carbohydrate metabolism in the two cell types.[11] The other kind of kinetic model describes and analyses the transition from metabolism at one point in time to another.

In order to construct a kinetic transition model of trehalose accumulation in _Dictyostelium_, it is necessary to measure the cellular levels of the relevant metabolites over the period of time to be analyzed, and to determine the flux of key reactions. The rates of reactions _in vivo_ may be determined using radioactive tracers, or may be calculated from metabolite concentration changes as a function of time.[8] Such data were obtained over a period of many years, and increasingly complex models were con-structed. These models of carbohydrate metabolism in _Dictyostelium_ eventually evolved into the metabolic network shown in Figure 7. This model consists primarily (90%) of data; the current model parameters are summarized below. (For references, see Wright and Kelly.[8]):
1.   Twenty-one flux patterns in the model. Five were deter-mined directly _in vivo_, with isotopes: The synthesis of trehalose, cellulose, and cell-wall glycogen, UDPG, glycogen, and the turnover of glycogen; these flux patterns were also consistent with values calculated from metabolite concentration changes during differentiation. Other flux patterns were based on the constancy or on the rate of change in metabolite concentrations (nucleotides, hexose and pentose phosphates, RNA, and glucosamino-glycan).
2.   Seven enzyme kinetic mechanisms and 27 kinetic constants for glycogen phosphorylase (R2), UDPG pyrophosphorylase (R4), trehalase (R6), glucokinase (R7), G6P phosphatase (R8), 5'-nucleotidase (R16), and uridine phosphorylase (R20).
3.   The assumption that the other reactions are adequately represented by mass action expressions; i.e., that substrate levels are far below K_m values.
4.   Three flux patterns for the entry of exogenous inorganic phosphate (P_i), glucose, and uracil over 900 or 90 minutes of differentiation. In order to simulate the results of the latter experiments, it was necessary to impose a precise (in timing and intensity) flux pattern of external metabolite on the model.
5.   Ninety metabolite accumulation patterns, which have been measured experimentally. Of these, 32 determinations were made

during normal differentiation: Glycogen, α- and β-cellulose, cell wall glycogen and trehalose, glucosaminoglycan, RNA, nucleotides, P_i, pentose and hexose phosphates, glucose, and UDPG. The other accumulation patterns were determined under specific conditions of perturbation by various exogenous metabolites for 900 minutes (15 determinations), or for 90 minutes (45 determinations), 20 of which are as yet unpublished.

6. The last parameter consists of 18 enzyme activation functions, $V_v(t)$ or $K_v(t)$, which are calculated, and do not change when the model is perturbed by external metabolites. This latter parameter requires further explanation.

Two kinds of rate constants are calculated: (a) one is used in a simple mass action expression (rate constant X metabolite concentration), when the substrate level is much less than K_m; i.e., the reaction is first order; and (b) the other is used in an expression for a specific enzyme kinetic mechanism. In this case, the rate constant is comparable to V_{max}; however, it is not determined in vitro but is calculated as the only unknown, based on the other parameters in the rate expression; i.e., the rate of the reaction, metabolite levels, and enzyme kinetic constants. We do not find it advisable to use enzyme specific activity as a model parameter, because enzyme activity in crude extracts rarely reflects activity in the intact cell or organism. Artifacts result from preparing cell extracts; e.g., enzyme activators or inhibitors may be diluted out, concentrated, lost, or created; compartmentation at the level of organelles, enzyme complexes, or enzyme-substrate complexes may be destroyed. Finally, enzymes usually are assayed under nonphysiological conditions of temperature, pH, substrate, effector, and protein concentration. Indeed, in our experience, there is a very poor correlation between enzyme activities in vivo and in vitro.

Ninety percent of the model parameters represent data; only the rate constants are calculated, based on these data. The ultimate test of any model is its predictive value, and over the years our models have made many predictions which have been tested subsequently by us or by others. As new data became available the model would be expanded and new predictions made. At least 50 predictions (some untestable) have been made, and about half of these have been examined experimentally. Most have been substantiated, or led to new predictions, especially concerning the compartmentation of metabolites. Other predictions have been concerned with the flux and turnover of metabolites in vivo, the activity patterns of enzymes both in vivo and in vitro, the enzyme kinetic mechanisms and constants of specific reactions, the presence of enzyme inhibition in vivo, the effects of external metabolites or altered enzyme activities on endogenous metabolite accumulation patterns, permeability patterns and compartmentation of enzymes.

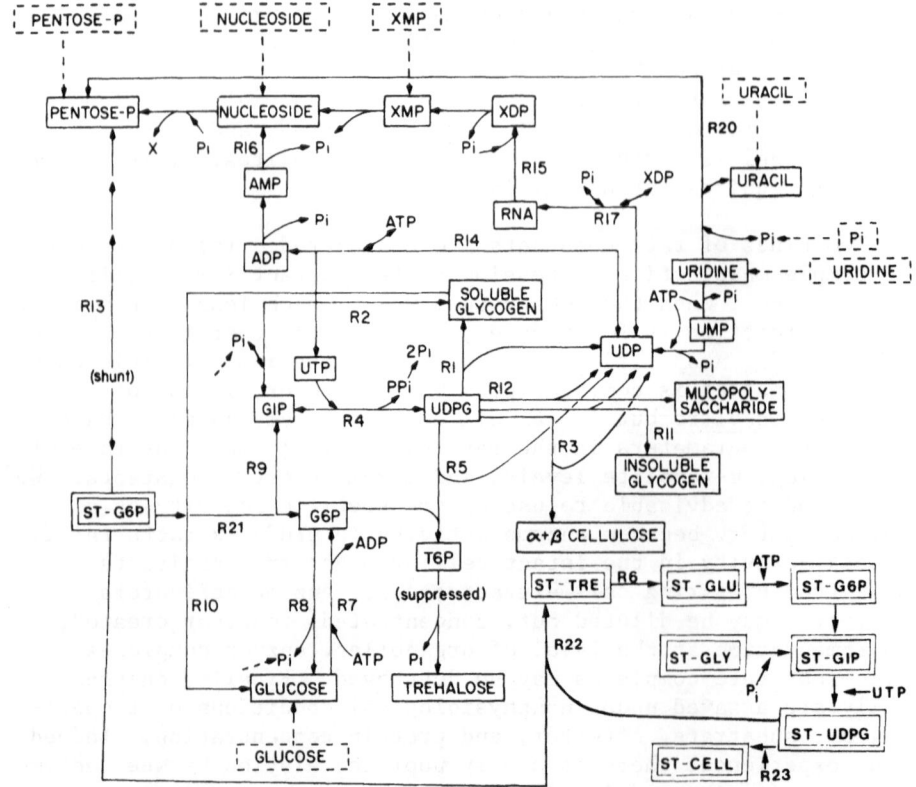

Fig. 7. The metabolic pathways and reactions representing a
kinetic model of carbohydrate metabolism in
Dictyostelium discoideum. Abbreviations: Pi, inorganic
phosphate; PPi, inorganic pyrophosphate; UTP, uridine
triphosphate; GlP, glucose 1-phosphate; G6P, glucose
6-phosphate; ATP, adenosine triphosphate; T6P, trehalose
6-phosphate; UMP, uridine monophosphate; UDP, uridine
diphosphate; UDPG, uridine diphosphoglucose; ADP,
adenosine diphosphate; AMP, adenosine monophosphate; XMP
and XDP refer to a mixture of the mono and diphosphates
of the bases in RNA. The double boxes represent stalk
(ST-) specific metabolites. The metabolites in broken
boxes are those used to perturb the organism and the
model.

Returning now to trehalose synthesis in this system: Did our model help us understand the rate-limiting events controlling the accumulation of this disaccharide? Of course the model represents an integrated system, with all pathways interacting and interdependent. However, I shall try to summarize a number of predictions and insights especially relevant to trehalose accumulation:

1. In vitro evidence indicated that the specific activity of trehalose-6-phosphate (T6P) synthase increased from aggregation to a maximum at culmination (catalyzing R5, Figure 1). However, simulations predicted that, because G6P and UDPG accumulated to maximum levels at culmination, this enzyme could not increase similarly in activity in vivo, as such an increase would prevent the accumulation of the two substrates. More than a year after this prediction was made, it was demonstrated that observed changes in vitro in the activity of the T6P synthase do not reflect activity in vivo: Using [^{14}C]glucose, the rate of trehalose synthesis was determined, based on the specific radioactivity of its precursors. This rate was negligible from aggregation until late in the culmination process, at which time a 100-fold burst in the rate of synthesis occurred.[12] It was also demonstrated that T6P synthase was "masked" in vitro and that artifacts were present that interfered with the assessment of enzyme activity in crude extracts.[12]

2. It was predicted that increasing the activity of glycogen phosphorylase during differentiation would enhance the levels of trehalose, G6P, and UDPG, in that order, but have little effect on cellulose levels. This prediction was unknowingly substantiated by Hames and Ashworth, who determined the levels of glucose, G6P, UDPG, trehalose, and cellulose in an axenic mutant with higher levels of glycogen phosphorylase.[13]

3. From observations of model behavior following perturbation by a flux of glucose, it was predicted that glucose perturbation of Dictyostelium should result in an increase in the trehalose/cellulose ratio. This prediction was substantiated.[8]

4. It is apparent from the kinetic constants for T6P synthase and the cellular concentrations of UDPG and G6P, that the latter substrate is in all probability the most rate-limiting in vivo.[14] Analysis of the model suggested that uracil should inhibit trehalose synthesis by lowering G6P levels, primarily by utilizing pentose phosphate (a source of G6P) to form uridine (R18). Based on the kinetic constants for glucokinase, G6P phosphatase and uridine phosphorylase, glucose should overcome this inhibition by increasing G6P levels via R7. Inorganic phosphate should enhance G6P accumulation and should overcome uracil inhibition by stimulating R18 (and hence R13) and by inhibiting R8.[15] These predictions were substantiated.[8]

5. Our dynamic understanding of the metabolic network leading to trehalose accumulation indicated that T6P synthase does not limit the rate of trehalose synthesis or the cessation of trehalose accumulation. This prediction was substantiated by the

observation that exogenous glucose could enhance trehalose levels
5-fold in mature (30 hour) sorocarps, many hours after the maximum
levels of trehalose had accumulated.[15]   Thus, glucose equivalents
are not only rate-limiting during differentiation, but are also
critical variables in the cessation of biochemical differentiation.

Most of the above predictions and insights into the control
of trehalose synthesis and accumulation resulted from the construction and analysis of kinetic models of this system.   Although the
purpose of these analyses was simply to understand the mechanisms
involved, this understanding could be used to enhance the levels
of trehalose.   We may then end this story with one more prediction:
The most efficient conversion of exogenous glucose to trehalose
should occur in aged sorocarps in the presence of high levels of
P_i.   Although P_i accumulates during development,[17] it does so in
stalk cells, which would no longer contain trehalose or be
metabolically active.   In the spore cells, however, G6P and
trehalose levels would be enhanced by the stimulation of R7 by
glucose, by the stimulation of R18 (R13) by P_i and by the inhibition of R8 (and perhaps R4) by P_i.   Glucose added at this time,
over relatively brief (e.g., 1 hour) periods, should primarily be
converted to trehalose; glycogen phosphorylase is undetectable,[16]
and glycogen turnover should have decreased to a very low level.

REFERENCES

1.   N.B. Madsen, The biological control of glycogen metabolism in
     Agrobacterium tumefaciens, Can. J. Biochem. Physiol. 41:561
     (1963).
2.   N.O. Sonza and A.D. Panek, Localization of trehalase and
     trehalose in yeast cells, Arch. Biochem. Biophys. 125:22
     (1968).
3.   C.P. Kubicek and M. Rohr, Influence of manganese on enzyme
     synthesis and citric acid accumulation in Aspergillus niger,
     Eur. J. Appl. Microbiol. 4:167 (1977).
4.   F.V. Soltero and M.J. Johnson, The effect of carbohydrate
     nutrition on penicillin production by Penicillium chrysogenum
     Q-176, Appl. Microbiol. 1:52 (1953).
5.   R.J. Neff and R.H. Neff, The biochemistry of amoebic encyst-
     ment, Symp. Soc. exp. Biol. 23:51 (1969).
6.   P.A. Rosness and B.E. Wright, In vivo changes of cellulose,
     trehalose and glycogen during differentiation, Archiv.
     Biochem. Biophys. 164:60 (1974).
7.   K.A. Killick and B.E. Wright, Trehalose synthesis during
     differentiation in Dictyostelium discoideum, Archiv. Biochem.
     Biophys. 170:634 (1975).
8.   B.E. Wright and P.J. Kelly, Kinetic models of metabolism in
     intact cells, tissues and organisms, Curr. Topics Cell.
     Regul. 19:103 (1981).

9. B.E. Wright, Modelling the environment for gene expression, in: "Gene Manipulations in Fungi," J.W. Bennett and L.L. Lasure, eds., Academic Press, N.Y. (in press).

10. P.J. Kelly, K.J. Kelleher, and B.E. Wright, The tricarboxylic acid cycle in Dictyolstelium discoideum. A model of the cycle at preculmination and aggregation, Biochem. J. 184:589 (1979).

11. B.E. Wright, D.A. Thomas, and D.J. Ingalls, Metabolic compartments in Dictyostelium discoideum, J. Biol. Chem. 257:7587 (1982).

12. K.A. Killick and B.E. Wright, Regulation of enzyme activity during differentiation in Dictyostelium discoideum, Ann. Rev. Microbiol. 28:139 (1974).

13. B.E. Wright and D.J.M. Park, An analysis of the kinetic positions held by five enzymes of carbohydrate metabolism in Dictyostelium discoideum, J. Biol. Chem. 250:2219 (1975).

14. K.A. Killick, Trehalose-6-phosphate synthase from Dictyostelium discoideum: partial purification and characterization of the enzyme from young sorocarps, Archiv. Biochem. Biophys. 196:121 (1979).

15. B.E. Wright, A. Tai, K.A. Killick and D.A. Thomas, The effects of exogenous glucose, uracil, and inorganic phosphate on differentiation in Dictyostelium discoideum, Archiv. Biochem. Biophys. 192:489 (1979).

16. D.A. Thomas and B.E. Wright, Glycogen phosphorylase in Dictyostelium discoideum. II. Synthesis and degradation during differentiation, J. Biol. Chem. 251:1258 (1976).

17. C.L. Rutherford, Cell specific events occurring during development, J. Embryol. exp. Morph. 35:335 (1976).

ANTIBIOTIC BIOSYNTHESIS - RELATION WITH PRIMARY METABOLISM

Giancarlo Lancini

Lepetit Research Laboratories
Via Durando 38
Milano

BIOGENESIS OF ANTIBIOTICS - GENERAL CONCEPTS

From the biogenetic point of view, antibiotics are secondary metabolites, a term widely used to indicate metabolites produced by plants or microorganisms which do not have any obvious function in the cell's growth, as demonstrated by the fact that the production is species or strain-specific and that mutants which have lost the ability to produce them are frequently found and have no impairment of vital function.
The study of their biogenesis consists of identifying the series of enzymatic reactions by which they are made, starting from one of the cellular "primary metabolites", that is, a normal cell constituent. Biosynthetic pathways can thus be ordered according to the primary metabolites from which they originate (W. Kurylowicz, 1976). This criterion, adopted with some elaboration by Turner in his "Fungal metabolites" (Turner, 1971), is however unsufficient when used by itself, since several different and unrelated pathways may take off from the same metabolite. Lancini and Parenti (1982) combined it with a somewhat formal classification of biosynthetic pathways, classifying them into three major classes : those starting from a primary metabolite, those in which two or more different metabolites are condensed and those involving polymerization processes.
A more natural and meaningful classification might be constructed on the assumption that the enzymatic reactions involved in antibiotic biosynthesis originate by modifications of the cells normal biochemical reactions.

With this as the basis, an attempt could be made to relate each antibiotic biosynthetic pathway to one of the
following classical biochemical processes, thus producing a classification scheme :

1) Glucose catabolism including : a) Embden-Meyeroff
   pathway; b) the pentose pathway and c) the tricarboxylic acid cycle.
2) Biosynthesis and the metabolism of amino acids.
3) Biosynthesis and the metabolism of the nucleotides.
4) Biosynthesis of the coenzymes.
5) Biosynthesis of fatty acids and membrane lipids.
6) Isoprene synthesis of branched aliphatic chains and
   sterols.
7) Synthesis of cell wall components, including :
   a) Conversion of glucose into other sugar and amino
      sugars;
   b) Oligomerization or polymerization of sugar;
   c) Other reactions such as peptidoglycan cross-
      linking, formation of glycolipids, etc.
8) Template directed synthesis of macromolecules :
   deoxyribonucleic acid, ribonucleic acids and proteins.

For each of these subdivisions several variants could
give origin to different families of antibiotics.

A tentative classification of the major recognized patterns of antibiotic synthesis based on this concept is
presented in this review. No attempt at completeness
will be made and for each group only one or a few examples will be given. One difficulty in this approach is
the presence in some antibiotics of chemical groups not
found in primary metabolites, such as the nitro group,
conjugated acetylene bonds, carbon-bonded halogens.
Although these groups are often essential for the antimicrobial activity, the enzymatic reactions that generate them can be considered to be secondary modifications
of the molecular structure and need not prevent the identification of its main pattern of biosynthesis. For the
same reason we can disregard such late modifications as
acylation, glycosylation, etc., which are also very important for the antimicrobial activity but rather unchar-
acteristic from the biosynthetic point of view. Although
for some cases attribution to a group is straightforward,
in many cases antibiotic appear to be made up of two or
more moieties derived from different biosynthetic pathways. In these cases, the moiety that appears to have
originated in the more typical biosynthetic reactions
has been considered more relevant to the classification.

BIOSYNTHETIC PATHWAYS RELATED TO GLUCOSE CATABOLISM AND THE TRICARBOXYLIC ACID (TCA) CYCLE

1) Glycolytic and pentose pathway

No important antibiotic has a structure derived from a series of reactions that can be considered to be variants of the glycolytic or pentose pathways, possibly because these pathways are catabolic rather than biosynthetic.
It is however obvious that the final products of these can be used as basic material in many biosynthesis : acetic acid gives rise to aliphatic chains through polymerization mechanisms, erythrose and pyruvic acid condense to give aromatic metabolites and ribose is present in nucleoside antibiotics and in several aminoglycosides. Occasionally intermediates between dihydroxyacetone and pyruvic acid can function as the starting material for part of complex molecules. It has been suggested, for instance, that methylglyoxal derived from dihydroxyacetone is the starter unit for the polyketide chain that originates the macrolide aplasmomycin (Chen et al., 1981). Another example is the glycol moiety of the ansamycin rifamycin B, which seems to originate from degradation of phosphoglyceric or phosphohydroxypyruvic acid (Lancini and Grandi, 1981).

2) Tricarboxylic acid cycle

The series of reactions of the tricarboxylic acid cycle also are catabolic rather than biosynthetic. Intermediates of the TCA cycle, such as α-ketoglutaric, oxalacetic and succinic acids, are the starting materials for important biosynthetic processes of both primary and secondary metabolism. Turner for instance (1971) lists several fungal metabolites derived from the condensation of TCA cycle intermediates, generally a four carbon dicarboxylic acid with a fatty acid. These metabolites, characterized by a five member lactone ring, are currently termed tetronic acids. Itaconic acid, although not an antibiotic, is a typical example of a secondary metabolite derived from a variant of a TCA cycle reaction. It originates from decarboxylation of cis-aconitic acid, an intermediate in the conversion of citrate to isocitrate (Bentley and Thiessen, 1957).

PATHWAYS RELATED TO AMINO ACID METABOLISM

The structure of many antibiotics or of parts of complex antibiotic molecules appear to be modifications of the structures of the twenty fundamental aminoacids. Available data however are often insufficient to establish whether these antibiotics originate in a variation of amino acid biosynthesis or in modification of a preformed amino acid.
One question not yet completely solved is, for instance, how the D-isomers of natural aminoacids, often found as constituents of peptide antibiotics, are biosynthesized. This matter has been reviewed by Katz and Demain (1977): for a few cases there is evidence that the D-isomer originated from the L-isomer and that the inversion takes place during the polymerization process, but this is far from being a general rule. In this section some pathways related either to aminoacid biosynthesis or to amino acid metabolism are examined. The assembly of aminoacids to give polypeptides will be discussed in the sections on oligomer formation and polymerization processes.

1) Variants of amino acid synthesis

Aromatic amino acid biosynthesis consists of an initial series of reactions, from erythrose phosphate to chorismate, common to all the final products and of specific branches leading from chorismate to tryptophan through anthranilic acid or to phenylalanine and tyrosine through prephenic acid.
The biosynthesis of several antibiotics can be considered to be related to these pathways.

a) Variants of the initial reactions.

In recent years, evidence has been accumulating about intermediates in antibiotic synthesis that originate in reactions diverging from the early steps of the shikimate pathway.
The best known example is that of 3-amino-5-hydroxybenzoic acid.
An aromatic unit characterized by an amino group meta to a carboxylic group (a combination unknown in primary metabolism) was postulated as the starting molecule for the poliketide chain that originates rifamycins and other ansamycins (White et al., 1973).
It was later shown that this moiety is derived

from an early intermediate on the shikimate pathway (White and Martinelli, 1974, Ghisalba and Nüesch, 1981). Besides constituting the chromophore of both benzene and naphthalene ansamycins, it is a major component of other antibiotics, such as mitomycin (Anderson et al., 1980, Hornemann, 1974).
A related compound, meta-aminobenzoic acid, is an intermediate in biosynthesis of the antibiotic pactamycin (Weller and Rinehart, 1978) and slightly different biosynthetic pathways have been proposed for both molecules (Rinehart et al., 1982).

b) Tryptophan branch.

Key steps in tryptophan biosynthesis are the condensation of anthranilic acid with phosphate-activated ribose and the subsequent cyclization to form the pyrrole ring.
The biosynthesis of the quinoline moiety of the antibiotic streptonigrin proceeds, according to recent studies (Gould and Cane, 1982), through reactions reminiscent of this process, consisting of the condensation of 4-aminoanthranilic acid with erythrose phosphate and subsequent cyclization to give the pyridine ring.

c) Phenylalanine tyrosine branch.

A few examples are known of variants of phenylalanine biosynthesis. The antibiotics 3,5-dihydrophenylalanine and ketomycin (cyclohexenylpyruvic acid) were both shown to be derived from shikimate and in both cases the producing strain was unable to utilize phenylalanine as precursor for their synthesis (Scannel et al., 1970), (Takeda and Floss, 1979). It is reasonable to assume that prephenic acid is the common precursor for these compounds.
Prephenic acid is most probably also the last common intermediate in the synthesis of p-aminophenylalanine. This unusual aminoacid has never been isolated in free form but is a key intermediate in the syntheses of chloramphenicol (McGrath et al., 1968) and of aureotine (Cardillo et al., 1972). It is noteworthy that in both cases it is first converted to p-aminophenylserine and that eventually the aromatic amine is oxidized to a nitro group.

2) Variants of amino acid catabolism

a) Aliphatic amino acids.

Branched chain amino acids. Valine, leucine and
isoleucine are degraded to small reutilizable molecules by oxidative pathways whose three initial
steps are identical. The amino acids are first
converted to the corresponding α-ketoacids by
transamination. An oxidative decarboxylation follows, yieding a branched aliphatic acid which is
then dehydrogenated.
Further reactions yield methylmalonyl-CoA from
valine, acetoacetyl-CoA from leucine and acetyl-
CoA and propionyl-CoA from isoleucine.
It is possible that the propionate or methylmalonate units necessary for the formation of the
macrolide, ansamycin and polyether antibiotics are
derived from valine and isoleucine through these
reactions. It may be surmised that these building
blocks are made by degradation of these amino acids
or from glucose by the glycolytic-Krebs cycle pathway, depending on the availability of the starting
material.
α-Hydroxy or α-keto acids are encountered in the
biosynthesis of depsipeptide antibiotics. It is
reasonable to assume that these are derived from
the initial steps of the branched amino acid catabolic pathway. There is evidence for the D-hydroxy-
isovaleryl moiety that constitutes with L-valine,
D-valine and lactic acid the depsipeptide ionopho-
re valinomycin. There is good evidence that D-hy-
droxyisovalerate and D-valine are both derived
from L-valine (Ristow et al., 1974). The lactyl
portion is labelled by threonine, which suggests
that it is derived from this aminoacid through the
catabolic pathway involving aminoacetone (Green,
1964) as an intermediate. D-hydroxyisovaleric
acid is also a constituent of the antibiotic
enniatin B and in this case too it originates
from valine (Zocker and Kleinhauf, 1978).
Another example is the α-keto-β-methylvaleric acid
portion of the peptide antibiotic Pyridomycin,
which has been shown to be specifically labelled
by L-isoleucine (Ogawara et al., 1968).

Lysine. A pathway of lysine catabolism which eventually gives acetoacetate through aminoadipic acid
involves as an intermediate piperidine -2- carboxy-
late.

Radiotracer experiments have shown that L-lysine
is the precursor of the 3-hydroxypicolinic acid
portion of the peptide antibiotic etamycin. It is
likely that 3-hydroxypicolinic acid is the result
of piperidine-2-carboxylate dehydrogenation (Hook
and Vining, 1973).

Heterocyclic rings. Heterocyclic rings originating from cyclization of aminoacids are found in
several antibiotics.
It is difficult to relate them to known aspects of
amino acid metabolism, in part for lack of sufficient information. A recently studied case is
that of the streptolidine moiety of Streptothricins
(Gould et al., 1981), which is derived from arginine. Another case is the azetidine moiety of the
nucleoside polyoxins, which originates from isoleucine (Isono et al., 1975).
Both in monobactam antibiotics and in nocardicin A,
the β-lactam ring carbons originate from serine.
It appears that the ring is formed by a simple
nucleophilic displacement of the activated seryl
hydroxyl by the serine carboxyl amide (Towsend
and Brown, 1981, O'Sullivan et al., 1982).

b) Aromatic amino acids.

Phenylalanine and tyrosine. The main pathways of
phenylalanine and tyrosine catabolism yield benzoic
and homogentisic acids, which can be further degraded to acetic acid or to dicarboxylic acids of
the TCA cycle. In both cases, the initial reaction
is transamination, to phenylpyruvic or to p-hydroxy-
phenyl pyruvic acid.
A second path of tyrosine catabolism that eventually yields melanins consists of hydroxylation to
dihydroxyphenylalanine (DOPA) and further reactions
in which the nitrogen is retained to form an imid-
azole ring.
An example of antibiotic biosynthesis related to
the first pathway is that of vancomycin. The two
p-hydroxyglycine residues in this antibiotic have
been shown to be derived from tyrosine. The conversion probably involves hydroxylation to β-hy-
droxytyrosine followed by transamination and oxidative decarboxylation. The β-hydroxylation of
tyrosine is also required for the formation of
another moiety of vancomycin, m-chloro-β-hydroxy-
tyrosine (Hammond et al., 1982).

The antibiotic novobiocin is composed of two aromatic portions and a modified sugar. The aromatic portions are both derived from tyrosine. One is a coumarin, made by simple cyclization of the alanyl chain. The other is a benzoic acid derivative, whose biosynthesis includes the degradation of the alanyl chain by the reactions above mentioned for formation of homogentisic acid. Radiotracer studies of lincomycin biosynthesis revealed that the propylhygric portion (N-methyl-4-propylproline) of this antibiotic originates from tyrosine. The labelling distribution is consistent with a bioconversion through DOPA and dopaquinone similar to that of the melanin biosynthetic patway (Witz et al., 1971).
The possible mechanisms of this aromatic ring fission have been discussed by Hurley and Speedie (1981), in connection with their biosynthetic studies of the antitumor agents anthramycin, tomaymycin, sibiromycin. These pyrrolobenzodiazepine antibiotics contain moieties which are closely related to the lincomycin proline derivative and they have also been shown to be derived from tyrosine.
In all the cases, the pyrrole or pyrrolidine rings originate from cyclization of the tyrosine alanyl chain, and degradation of the aromatic ring provides the atoms of the aliphatic side chain.

Tryptophan. The main pathways of tryptophan catabolism initiate with the oxidative opening of the pyrrole ring to form kynurenine, which can be degraded to anthranylates or, by transamination and ring closure, can give knurenate.
The above mentioned pyrrolobenzodiazepine antibiotics are made by condensation of two units, the pyrrole ring derived from tyrosine and an anthranilic acid derivate. A detailed study (Hurley and Gairola, 1979) has shown that the anthranilic moiety derives from tryptophan through the kynurenine pathway.
The chromophore of actinomycin D also contains an anthranilic acid moiety derived from tryptophan. Evidence that it is also derived from kynurenine degradation has been reported (Perlman, 1973).
Tryptophan is the percursor of the antifungal antibiotic pyrrolnitrin, one of the few examples of a nitro-containing natural product. The biosynthe-

82

šis of pyrrolnitrin also involves the opening of the tryptophan pyrrole ring, with retention of the amino group but by a mechanism that differs from that of kynurenine synthesis. It has been suggested that a recently isolate enzyme, indolyl alkane hydroxylase catalyzes this reaction (Chang et al., 1981).

PATHWAYS RELATED TO NUCLEOTIDE METABOLISM

About seventy nucleoside antibiotics have been reported so far in the literature.
Their isolation, structure, chemical and biological properties have been recently reviewed (Suhadolnik, 1979).
A review specifically discussing their biosynthesis has also recently been published (Suhadolnik, 1981).
Biosynthetic information is available for about twenty compounds, and in the majority of cases their biosynthesis is, as expected, either related to that of natural nucleotides and deoxynucleotide or consists of rather simple modifications of nucleosides commonly present in the cells.

1) Variants of nucleotide synthesis

There is an interesting example of an antibiotic whose biosynthesis closely parallels the "de novo" synthesis of pirimidine nucleotides. This is 5,6-dihydro-5-azadeoxythymidine, a simmetrical triazene deoxyriboside. The biosynthetic sequence starts from glyoxyl urea (formed by condensation of urea with glioxilic acid) instead of aspartic acid and, as in uridine biosynthesis, carbamoyl phosphate supplies the additional carbon and nitrogen necessary for ring closure. The subsequent reactions, riboxylation, decarboxylation, etc. are analogous to those in normal nucleotide biosynthesis (Suhadolnik, 1981).
The "de novo" synthesis of deoxyribonucleotides consists of the reduction, catalysed by ribonucleotide reductase, of the hydroxyl at C-2' of the corresponding ribonucleotides. Cordicepin or 3-deoxyadenosine is an antibiotic synthesized by a modification of this process. In fact it has been shown that not only is it derived from adenosine (Chassy and Suhadolnik, 1969) but also that the reduction mechanism at C-3' of the latter is similar to that of the reduction at

C-2' in deoxynucleotide synthesis (Lennon and
Suhadolnik, 1976). Microorganisms can synthesise
nucleotides by the socalled salvage pathways, consisting of the ribosylation of a preformed purine or
pyrimidine base. Psicofuranine is an analog of adenosine in which the hexose psicose substitutes for ribose.
Evidence has been reported (Sugimori and Suhadolnik,
1965) showing that it is not derived through a modification of adenosine ribose but through the addition
of the sugar psicose to free adenine.
Its biosynthesis can thus be considered a variation
of a salvage pathway.
A similar pathway is discernible in the biosynthesis
of the antitumor agent blasticidin S. This antibiotic has a nucleoside moiety, cytosinine, consisting of
cytosine and an amino-hexuronic acid. Cytosinine formation proceeds through condensation of glucose with
cytosine to subsequent oxidation of the sugar moiety
to glucuronic acid (Seto et al., 1968), (Seto et al.,
1976).

2) Nucleotides modifications

Several nucleoside antibiotics consist of normal
purine or pyrimidine bases bearing unusual sugar moieties instead of ribose. The sugar can carry one or more
aminoacids as substituents. These antibiotics can be
made either by a modification of nucleotide synthesis,
as discussed in the previous section, or by biosynthetic modifications of the corresponding preformed
nucleotides or nucleosides.
Adenosine is for instance the direct precursor of
several antibiotics. A simple case is that of Ara-A
(arabinosyl adenine) biosynthesis, in which adenosine
is directly converted into the final product by the
enzyme 2'-adenosyl epimerase (Farmer et al., 1973)
(Suhadolnik, 1981).
Similarly, it has been shown that adenosine is the
precursor of the cytostatic antibiotic 3'-amino-3'-
deoxyadenosine.
A 3'-amino-3'deoxyadenosine moiety, also derived from
adenosine, is present in the protein synthesis inhibitor puromycin (Suhadolnik, 1981). The biosynthesis
of this antibiotic involves the N-methylation of this
moiety, addition of tyrosine and O-methylation of the
latter. Little is known about order of these reactions except that the O-methylation appears to be the
last step in the sequence (Sankaran and Pogell, 1975).

PATHWAYS RELATED TO COENZYME BIOSYNTHESIS

Coenzymes are molecules that differ widely in their chemical structures, reflecting the diverse biochemical functions they perform. They are made by a variety of biosynthetic pathways, some of which have not yet been studied throughly. In several cases, however, it is possible to relate the biosynthesis of an antibiotic to that of a coenzyme. We report here examples related to biotin, to nicotinic acid, to the pteridine rings of riboflavin and folic acid, and to small peptides, such as glutathione and coenzyme A.

a) Biotin

A few antibiotics related structurally to biotin have been described. Biosynthetic information is available for α-dehydrobiotin only. Labelled biotin, but not the biotin precursor, pimelic acid is converted by the producing microorganism into α-dehydrobiotin.
This fact and the inability of the latter to support the growth of biotin-requiring microorganisms indicate that dehydrobiotin is biosynthesized from biotin (Hanka et al., 1969).

b) Nicotinic acid

Nicotinic acid is a precursor in the synthesis of nicotinamide-adenine-dinucleotide (Brown and Reynolds, 1963).
In animals and in Neurospora, it is biosynthesized from tryptophan through 3-hydroxyanthranilic acid, whereas in plants and in bacteria it results from the condensation of a C-4 dicarboxylic acid and a C-3 unit (Brown and Reynolds, 1963).
Two pyridine ring moieties are present in the peptide antibiotic pyridomycin. Incorporation studies have shown that both are derived from aspartate and a C-3 unit, possibly glyceraldehyde indicating that, unlike that of pyridine moiety of etamycin (Hook and Vining, 1973), their biosynthesis in bacteria parellels that of nicotinic acid. (Ogawara et al., 1968)

c) Riboflavin and folic acid

In both riboflavin and folic acid, there is a pteridine moiety which originates biosynthetically from GTP (Burg and Brown, 1968, Plaut et al., 1974).

The first reaction in the biosynthetic process is for
both cases, removal of GTP's carbon 8, as formate, by
the enzymes GTP-cyclohydrolase I and II (Foor and
Brown, 1980).
The structures of the nucleoside antibiotics Tuberci-
din, Toyocamycin and Sangivamycin are characterized
by a ribosylated pyrrolopyrimidine ring. There is
evidence that the carbons of the pyrimidine moiety
originate from GTP and that the first reaction in
their biosynthesis in the removal of carbon 8 of GTP,
as formate, by an enzyme called GTP-8-formylhydrolase
(Uematsu and Suhadolnik, 1970; Elstner and Suhadolnik,
1971).
The carbons for the pyrrole ring formation are provided by ribose, again paralleling folic acid biosynthesis, in which ribose contributes to the pteridine's
pyrazine ring.

d) Glutathione and coenzyme A

Most of the peptide antibiotics are synthesized by a
multienzyme thiotemplate mechanism that will be discussed in the next section. However, it appears
that small microbial oligopeptides, for instance
tripeptides, are made by stepwise condensation of the
constituent aminoacids, probably activated as the
phosphates. This mechanism is similar to that of
glutathione synthesis (A. Meister, S.S. Tate, 1976)
and to some aspects of coenzyme A assembly (Brown and
Reynolds, 1963; Plaut et al., 1974).
The most important example is undoubtedly the synthesis of the tripeptide L-αaminoadipyl-L-cysteyl-D-
valine, which is the backbone of the β-lactam antibiotics, penicillins, cephalosporins and cephamycins
(Fawcett et al., 1976). A glutathione-like mechanism
for biosynthesis of this tripeptide, involving stepwise assembly for the aminoacids, was first proposed
by Demain (1966) and evidence was provided by Loder
and Abraham (1971), who carried out cell-free experiments and showed that aminoadipyl-cysteine is first
formed and then condensed with valine.
The cyclization to form the β-lactam ring and the
further modifications to yield the final molecules
have been extensively studied. Among the several
reviews on this subject available, recent ones have
been published by Sullivan and Abraham, 1981, Hook et
al., 1982, Queener and Nuess, 1982.
Good evidence for a phosphate activated synthesis is
available for the microbial enzyme inhibitor leupeptin
(Suzukake et al., 1982).

Umezawa and coworkers have, in fact, isolated the enzyme leupeptin synthase (Suzukake et al., 1979), which catalyzes the condensation of acetyl-leucine and leucine to give acetyl-leucyl-leucine and the condensation of this dipeptide with arginine to give acetyl-leucyl-leucyl-arginine.

PATHWAYS RELATED TO FATTY ACIDS BIOSYNTHESIS

The biochemical process by which fatty acids are synthesized essentially consists of the stepwise head-to-tail condensation of acetate units activated by carboxylation to malonate and by thio-esterification to a carrier protein (Lynen 1967). The carbon skeleton of a number of antibiotics are biosynthesized by a closely related mechanism (Lynen and Tada, 1961). A partial modification of the function of the synthesizing enzyme or the utilization of building blocks other than acetate-malonate can generate the following classes of secondary metabolites :

1) Polyketide-derived aromatic molecules
2) Macrocylic lactones
3) Polyether antibiotics
4) Ansamycins

In addition, most of the peptide antibiotics are biosynthesized by the so-called multienzyme thiotemplate mechanism of amino acid polimerization. This mechanism also has strong analogy to the basic process of fatty acid synthesis (Lipmann, 1971) and is therfore discussed in this section.

1) Polyketide synthesis of aromatic molecules

When during the process of fatty acid synthesis the addition of a malonate unit to the growing chain is not followed by the NaDPH-dependent reduction of the β-carbonyl group, the result is a polyketomethylene (or polyketide) chain, that is an aliphatic chain in wich keto groups alternate with methylenes. Such a system, which is chemically very reactive, tends to fold and give aromatic rings through internal aldol condensation.
The structure of the final compound produced depends on the length of the chain, on the way it cyclizes and on the further enzymatic reactions for which the molecule may be a substrate.

Demonstration that this mechanism of polymerization operates in synthesis of an antibiotic is generally based on the incorporation of suitable labelled precursors into the final molecule. Evidence that the similarity between the processes of primary and the secondary metabolism is not only a formal one was provided by studies of 6-methyl-salicilic acid biosynthesis. An enzyme, 6-methylsalicilic acid synthetase, has been isolated and shown to be structurally very similar to although not functionally identical with the fatty acid synthetase of the same organism, Penicillium patulum. It may be assumed that the two enzymes share a common phylogenetic origin (Dimroth et al., 1970, Behal, 1982).

Turner, in his book on fungal metabolites lists over three hundred aromatic compounds. He proposed the terms triketide, tetraketide, etc. to denote products derived from three, four, etc. acetate-malonate units. The biosynthesis of several polyketide-derived antibiotics has been reviewed (Tanenbaum, 1967, Floss, 1981).

Extensive studies have been reported for the tetracyclines, the antitumor anthracyclines and the antifungal griseofulvin.

Tetracyclines can be considered to be decaketides originating from a hypothetical chain of which malonamide is the starter unit and nine malonates have been added in elongation.

The structure of the final product, tetracycline, oxytetracycline, chlortetracycline is the result of a sequence of complex reactions that greatly modify the initial molecule. However, the first known intermediate in the biosynthesis, 6-methyl-pretetramide, consisting of four aromatic rings, clearly possesses the chemical functions expected from the hypothesised origin (McCormick, 1967, Vanek and Hostaleck, 1973, Hutchinson, 1981).

Griseofulvin, an antifungal antibiotic, belongs to the large group of heptaketides produced by fungi. The first intermediate in its biosynthesis has the aromatic structure of a substituted benzophenone. The aromatic character, as in the case of tetracyclines, is partially lost in the subsequent modifications (Grove, 1967, Turner, 1975).

The anthracycline antibiotics, which include several antitumor agents, are glycosyl derivatives of decaketide aglycones, called anthracyclinones. Several studies of the biosynthesis of these antibiotics are available and a review on this subject has been recently published (Vaneck et al., 1982).

Incorporation of ^{13}C labelled precursors has given
support to the polyketide origin of the aglycones of
adriamycin, nogalamycin, steffimycin (Casey et al.,
1978, Paulick et al., 1976, Wiley et al., 1978) and
has provided information about the mode of cyclization.
A number of intermediate anthracyclinones have been
isolated from fermentations of several Streptomyces
strains. They can be considered derivatives of either
aklavinonic acid, the cyclization product of a polyketide chain initiated by propionate or of α and β-rho-
domycinone, in which acetate is the starter molecule
(Vaneck et al., 1976, 1982).

2) Macrocylic lactones

In fatty acid biosynthesis, the process of reduction of the growing chain consists of hydrogenation
of the β-carbonyl to a hydroxyl group followed by
dehydration and further hydrogenation of the resulting
double bond. If this process is only partially performed, the resulting chain bears either hydroxyl on
alternate carbon atoms or, if the dehydration step has
occurred, a series of conjugated double bonds.
This biogenetic mechanism can give rise to the macrolide antibiotics, in which the chain forms a cycle
closed by a lactone bond.
A characteristic of these antibiotics is the presence
on the chain on methyl groups derived from the utilization of propionate-methylmalonate instead of acetate-malonate as elongation units.
It has been suggested (Corcoran and Chick, 1966) that
in the biogenesis of these antibiotics the process of
reduction may be incomplete because these methyl groups
can sterically hinder the binding of the reducing
enzymes to the growing chain.

a) Macrocyclic polyenes

The typical structure of these antifungal antibiotics is a macrocyclic lactone of 26 to 38 atoms
making up a region with a series of conjugated double bonds, from four to seven.
Other usual structural features are a series of
hydroxyl groups and a sugar, mycosamine, glycosidi-
cally linked to the macrocycle. A compilation of
polyene antibiotics and of their producing strains
has been made by Pridham and Lyons (1976).
Their biochemical and biological aspects have been
reviewed by Hamilton Miller (1973), their production
by Martin and McDaniel (1977), their biosynthesis by

Martin (1977). Biosynthetic studies are available
for about ten polyene antibiotics; the overall evidence is in agreement with biogenetic origin from
a partially reduced polyketide chain and the evidence can be summarized as follows :

- acetate and propionate are efficiently incorporated into the macrocyclic ring of all the antibiotics tested (Martin, 1977).

- in the nystatin producer Streptomyces noursei,
  the enzymes that convert acetate to acetyl CoA
  and this to malonyl CoA are associated with antibiotic production.
  The same is true for the enzymes that convert
  propionate to methylmalonyl CoA (Martin 1977).

- Cerulenin, an inhibitor of fatty acid synthase,
  inhibits the synthesis of the polyene candicidin.
  This suggests that macrolide chain synthase is
  similar to fatty acid synthase (Martin and
  McDaniel, 1975).

b) Antibacterial macrolides

This group is characterized by a relatively smaller
size of the macrocylic lactone, with from 12 to 16
atoms. At least one glycosidic substituent is
always present. A general review of macrolides,
with emphasis on the chemical aspects, and a review
of the chemical and biological studies of 16-mem-
bered macrolide antibiotics have been published
(Masamune et al., 1977, Omura and Nakagawa, 1975).
Review articles dealing specifically with biosynthetic studies have been written by Vanek and
Mayer (1967) and Omura and Nakagawa (1981).
The most thoroughly studied macrolide biosynthesis
is that of erythromycin and an extensive review on
this subject has been published by Corcoran (1981).
The aglycone of erythromycins, erythronolide B is
synthesized by condensation of one propionate and
six methylmalonate units by a multienzyme complex,
a synthase with the same size as the synthase
involved in membrane fatty acid synthesis. Whether
the two enzymes are identical or only very similar
is still an open question : if they are identical,
the choice between synthesis of an antibiotic molecule and that of a fatty acid would depend on the
availability of the precursors, acetate-malonate
or propionate-methyl propionate respectively
(Corcoran, 1981).

The patterns of biogenesis of the other macrolides
are similar to that of erythromycin, but with some
differences in the building units.
Those with a 14 atom aglycone, such as picromycin,
narbomycin and oleandomycin are not exclusively
derived from propionate-methylmalonate but contain
one unit that originates from acetate-malonate
(Masamune et al., 1977).
Those with 16 membered rings invariably contain a
unit derived from butyrate-ethylmalonate and can
be divided into two groups : tylosin, rosamycin
etc., largely derived from propionate, and carbo-
mycin, leucomycins, spiramycin etc., mainly formed
from acetate-malonate (Omura and Nakagawa, 1981).

3) Ionophore polyether antibiotics

The typical structure of polyether antibiotics
is an aliphatic chain partially cyclized to saturated
pyran and furan rings by a series of ether linkages.
The chain ends with a carboxyl group and carries
methyl, ethyl and oxo-substituents. About fifty
polyether microbial metabolites have been described
(Westley, 1977).
Biosynthetic studies have been reported for five of
them : lasalocid, monensin, narasin, salinomycin, and
lysocellin. These have been recently reviewed
(Westley, 1981). In all the cases examined, incorporation of 13C-labelled precursors was consistent with
a mechanism of assembly similar to that in fatty acid
biosynthesis.
Typical of polyether biosynthesis is the frequent par-
ticipation of butyrate-ethyl malonate units in the
chain formation which results in the presence of ethyl
substituents on the final molecule.
Another interesting result of these studies is the
demonstration of previously unknown metabolic pathways
for conversion of butyrate to propionate.
The cyclic antibiotics aplasmomycin and boromycin, the
only boron-containing natural products, have some
characteristics in common with linear polyethers, particularly the saturated furan and pyran rings.
Aplasmomycin and boromycin appear to be biosynthesized
by condensation of two identical chains almost entire-
ly derived from malonate units.
An unusual feature is that there are several methyl
groups on the chain which are not derived from propionate-methyl malonate incorporation but from methionine
(Floss and Chang, 1981).

4) Ansamycins

This family of antibiotics is characterized by a macrocycle including an aliphatic chain and an aromatic nucleus. Unlike in macrolides, the cycle is closed by an amide bond on the aromatic chromophore. Depending on the structure of the latter, there are benzene ansamycin and naphtalene ansamycins, endowed with clearly different biological properties (Brufani, 1977, Lancini and Zanichelli, 1977). Most of the studies of ansamycin biosynthesis have dealt with the rifamycins, the streptovaricins and geldanamycin; a recent review has been published by Lancini and Grandi (1981). Labelled precursor incorporation (White et al., 1973) and isolation of intermediates from blocked mutants (White et al. 1974) have demonstrated that ansamycins originate from a partially reduced polyketide chain, initiated by an aromatic moiety and made up of malonate and methyl malonate as elongation units. The aromatic moiety, which, as mentioned before, is derived from the shikimate pathway, has recently been identified as 3-amino-5-hydroxybenzoic acid (Kibby et al., 1980, Ghisalba and Nuesch, 1981) and this is the nucleus of benzene ansamycins. In naphthalene ansamycins, it becomes part of the aromatic structure, the second ring being formed by partial cyclization of the aliphatic chain. Studies of the enzymes involved in ansamycin synthesis have not so far been reported. However, as pointed out by Prelog and coworkers (Brufani et al., 1973), portions of ansamycin and of the macrolide aliphatic chain not only have the same substitution pattern but also identical stereochemical configuration, which suggests a strong similarity of synthesizing enzymes.

5) Peptide and depsipeptide biosynthesis

About three hundred antibiotics with oligo or polypeptide structure have been reported in the literature. Their chemical characteristics have been summarized by Perlman and Bodanszky (1971) : they differ from proteins in their small size, rarely more than 2000 daltons, and in. the presence of both D- and L-aminoacid residues and of unusual aminoacids, and many have cyclic structures. The depsipeptide antibiotics, composed of both amino acid and hydroxy acid residues, in which peptide bond alternate with ester bonds, have similar characteristics.

Several recent reviews on peptide antibiotic biosynthesis are available (Lipmann, 1980, Kurahashi, 1974, Kurahashi, 1981).
A comprehensive review of the chemistry and biology of peptide antibiotics produced by bacilli has been published by Katz and Demain (1977).
As discussed in a previous section, the small peptides (tripeptides and perhaps tetrapeptides) are biosynthesized by a simple mechanism involving the activation of the amino acids as phosphates.
On the contrary, biosynthetic studies of larger antibiotics, with five or more residues, revealed a complex mechanism of assembly with the following characteristics (Lipmann, 1980) : amino acids are activated by ATP to aminoacyl adenylates, then transferred to thiol acceptors on a multienzyme complex, where they form thioester bonds.
The subsequent condensation process is mediated by 4-phosphopantetheine, which is attached through a carrier protein to the multienzyme system. The carboxy of the initiator aminoacid is transesterified to the pantetheine chain which transfer it onto the amino group of the succeding amino acid; the nascent peptide chain is again transesterified onto pantetheine and the process is repeated until completion of the peptide chain.
The sequence of the amino acids in the peptide is determined by the substrate specificity of the subunits of the synthetase and by the spatial arrangement of the thiol groups.
For this reason, the system is called the multienzyme thiotemplate mechanism.
Although the initial activation as acyl adenylates is reminiscent of protein synthesis, the chain formation mechanism is clearly very similar to that of polyketide synthesis, in which a multienzyme complex also directs the pantheine-mediated condensation of thioester-activated units. It is typical of both systems that incomplete chains do not serve as precursors for the final molecule.
The process has been studied in detail for the tyrocydins, gramicidin S and linear gramicidins. The synthetases for these antibiotics have been isolated and disaggregated into smaller subunits whose specific functions have been established (Kurahashi, 1981).
Less detailed information is available for the enzyme that synthesizes edeine, alamethacin, bacitracins, polymixins. The biosynthesis of the depsipeptide valinomycin and of enniatin B has also been studied.

There are indications that a thiotemplate multienzyme
mechanism operates in these cases too, but no firm
conclusion can be drawn at present.

ISOPRENOID BIOSYNTHETIC PATHWAY

Terpenoid substances are a class of metabolites with
rather diverse chemical structures that share a common
biogenetic origin.
This consists of the condensation of acetate units to
give isopentenyl-pyrophosphate, through the intermediate
mevalonic acid, and the subsequent oligomerization of
the isopentenyl units (Beytia and Porter, 1976).
Terpenoids are widespread among microorganism metabolites : for instance, fungi have sterols as essential
components of their cell membranes, isoprenoid quinones
are common in all bacteria (Collins and Jones, 1981) etc.
It is thus somewhat surprising to observe that only a
small number of antibiotics have structures indicative
of this biosynthetic origin.
Among streptomyces products, the clearly defined only
case is the sesquiterpene antibiotic pentalenolactone
(Areneamycin, PA 132).
Since exogenous acetate and mevalonate were not utilized
by the producing strain, its biosynthetic pattern was
analyzed from the incorporation of ^{13}C-labelled glucose
(Cane et al., 1981), detected by NMR spectroscop.
Pentalenolactone appears to be biogenetically related to
the humulene-derived fungal antibiotics, the illudins
(Hansen, 1976).
The biosynthesis of aphidicolin, a diterpene fungal metabolite of interest for its inhibitory effects on DNA
polymerase, was studied in some detail (Adams and
Bu'Lockn 1975). The arrangement of the constituent isoprene units has been defined, as well as the rearrange-
ment steps during biosynthesis (Ackland et al., 1982).
A few steroidal antibiotics have been described. These
include fusidic acid and its co-metabolites, helvolic
acid, and cephalosporin P_1.
The pattern of fusidic acid biosynthesis and its relation
to lanosterol biosynthesis were established by studies
of incorporation labelled acetate (Riisom et al., 1974),
mevalonic acid (Caspi Mulheirn, 1970) and squalene oxide
(Godtfredsen et al., 1968).
A protostane derivative, a common intermediate in the
biosynthesis of fusidic acid, helvolic acid and cephalosporin P_1, was synthesized from mevalonate by cell-free
extracts of a helvolic acid producer (Kawaguchi, 1970).

PATHWAY RELATED TO POLYSACCHARIDE BIOSYNTHESIS

Microbial polysaccharides are also a diverse group
of biopolymers : they can roughly be divided into intracellular polysaccharides, mainly glucans with a reserve
function, and extracellular polysaccharides, mainly
heteropolysaccharides, including structural components
of cell walls and capsules and exopolysaccharides excreted into the culture medium (Woodside and Kwapinsky, 1974).
The pattern of their biosynthesis is, with some excep-
tions, rather uniform : glucose activated as phosphate
or as nucleotide-diphosphate is enzymatically transformed
into the sugars constituting the polysaccharide. The
nucleotide-diphosphate sugars are, as a rule, the activated units that take part in the polymerization process.
The oligosaccharide antibiotics resemble the structural
and exopolysaccharides in the variety of the monomers
that constitute them.
However, their size is comparatively smaller, rarely more
than four units.
Since aminosugars are constituents of most of the oligosaccharide antibiotics, and certainly of all the important ones, these are referred to as aminoglycoside antibiotics. Another term used is amino-cyclitol antibiotics, because of the substituted cyclitol (a polyhydroxy-
lated cyclohexane) in the majority of them. Their chemical and biological properties have been summarized in a
comprehensive monograph (Humezawa and Hooper ed. 1982).
Reviews specifically dealing with aminocyclitol antibiotic biosynthesis have recently appeared (Pearce and
Rinehart, 1981, Okuda and Ito, 1982).

a) Biosynthesis of the subunits

The origin of the subunits from glucose, without
carbon rearrengement, has been established for a large
number of cases, mainly by isotope labelling experiments.
In the case of streptidine, the aminocyclitol of
streptomycin, the intermediates in the conversion have
been identified and the enzymatic transformations in
a cell-free system described (Demain and Inamine, 1970,
Walker, 1975). Bluensomycin's cyclitol is biosynthesized by a similar pathway (Walker, 1974).
Myoinositol, found in a variety of organisms, is the
first intermediate in the reaction sequence. It is
also an intermediate in the formation of actinamine,
the aminocyclitol moiety of Spectinomycin (Mitscher
et al., 1971, Stroshane et al., 1976) and of D-chiro-
inositol the cyclitol of Kasugamycin (Fukagawa et al.,1968).

However deoxystreptamine, the aminocyclitol of most
aminoglycoside antibiotics, is biosynthesized by a
different route from that of streptidine.
Glucose is converted first to 2-deoxyscylloinosose,
as proposed by Rinehart and Stroshane (1976) and confirmed by Kakimura and coworkers (1981).
In the few cases studied, the conversion of glucose
to other sugars appears to involve, as in primary metabolism, its activation as the nucletidyl-diphosphate.
Grisebach and coworkers (Ortmann et al., 1974, Whal
et al., 1975) demonstrated that in streptomycin biosynthesis dTDP-glucose is converted into dTDP-dihydrostreptose, whereas in the formation of methyl-L-gluco-
samine the precursor, D-glucosamine, appears to be
activated as the uridine diphosphate (Hirose-Kumagai
et al., 1982).
Evidence for the involvment of UDP-acetylglucosamine
in the biosynthesis of kasugamine, the sugar Kasugamycin, was provided by Fukagawa et al., (1968).

b) Subunit assembly

The subunit assembly initiates in general with the
condensation of a sugar to the aminocyclitol moiety.
In streptomycin biosynthesis, dihydrostreptose, activated as dTDP-dihydrostreptose, is condensed to
streptidine-6-phosphate (Kniep and Grisebach, 1976).
There is evidence that in all deoxystreptamine-con-
taining antibiotics the sugar first added to the
cyclitol is D-glucosamine, to form the pseudodisac-
charide paromamine, however the details of this
reaction are not known (Okuda and Ito, 1982).

PATHWAYS RELATED TO NUCLEIC ACID AND PROTEIN BIOSYNTHESIS

Two antibiotics only, Nisin and Subtilin, appear to
be biosynthesized by the mechanisms of protein synthesis.
These antibiotics are relatively large peptides, with 34
and 32 aminoacid residues respectively.
The available evidence, reviewed by Kurahashi (1981), is
based on inhibition of nisin synthesis by inhibitors of
the m-RNA-ribosome system (Hurst, 1966, Ingram, 1970).
The same type of evidence has been given (Kurahashi,
1981) for subtilin synthesis. Both subtilin and nisin
seem to be derived from larger inactive precursor peptides
(Hurst and Paterson, 1971).

REFERENCES

M.J. Ackland, J.R. Hanson, A.H. Ratcliffe, I.H. Sadler, J. Chem. Soc. Chem. Commun., 165-166, 1982.

M.R. Adams, J.D. Bu'Lock, J. Chem. Soc. Chem. Commun.; 389-391, 1975.

M.G. Anderson, J.J. Kibby, R.W. Rickards, J.M. Rothschild, J. Chem. Soc. Res. Commun., 1277-1278, 1980.

V. Behal in "Overproduction of Microbial Products", V. Krumphanzl, B. Sikyta, Z. Vanek Eds., Academic Press London, 1982, pp. 301-309.

R. Bentley, C.P. Thiessen, J. Biol. Chem., 226, 673-687, 703-720, 1957.

E.D. Beytia, J.W. Porter, Ann. Rev. Biochem., 45, 113-142, 1976.

G.M. Brown, J.J. Reynolds, Ann. Rev. Biochem., 32, 419-462, 1983.

M. Brufani, D. Kluepfel, G.C. Lancini, J. Leitich, A.S. Mesentsev, V. Prelog, F.P. Schmook, P. Sensi, Helv. Chim. Acta, 56, 2315-2323, 1973.

M. Brufani in "Topics in antibiotic chemistry", Vol. I, P.G. Sammes Ed. Horwood Chichester, 1977, pp. 91-212.

A.W. Burg, G.M. Brown, J. Biol. Chem., 243, 2349-2358, 1968.

D.E. Cane, T. Rossi, A.M. Tillman, J.P. Pachlatko, J. Am. Chem. Soc., 103, 1838-1843, 1981.

R. Cardillo, C. Fuganti, D. Ghiringhelli, D. Giangrasso, P. Grasselli, Tetrahedron Letters, 4875-4878, 1972.

M.L. Casey, R.C. Paulick, H.W. Whitlock, J. Org. Chem., 43, 1627-1634, 1978.

E. Caspi, L.J. Mulheirn, J. Am. Chem. Soc., 92, 404-406, 1970.

C.J. Chang, H.G. Floss, D.J. Hook, J.A. Mabe, P.E. Manni, J. Antibiot., 34, 555-566, 1981.

B.M. Chassy, R.J. Suhadolnik, <u>Biochim. Biophys. Acta</u>, 182, 307-315, 1969.

T.S. Chen, C. Chang, H.G. Floss, <u>J. Am. Chem. Soc.</u>, 103, 4565-4568, 1981.

M.D. Collins, D. Jones, <u>Microbiol. Reviews</u>, 45, 316-354, 1981.

J.W. Corcoran, M. Chich in "Biosynthesis of Antibiotics", J.F. Snell Ed. Academic Press, New York, 1966, pp. 159-201.

A. Demain in "Biosynthesis of Antibiotics", J.F. Snell Ed. Academic Press New York, 1966, p. 29-94.

A.L. Demain, E. Inamine, <u>Bacteriol. Reviews</u>, 34, 1-19, 1970.

P. Dimroth, H. Walter, F. Lynen, <u>Eur. J. Bioch.</u>, 13, 98-110, 1970.

E.F. Elstner, R.J. Suhadolnik, <u>J. Biol. Chem.</u>, 246, 6973-6981, 1971.

P.B. Farmer, T. Uematsu, H.P.C. Hogenkamp, R.J. Suhadolnik, <u>J. Biol. Chem.</u>, 248, 1844-1847, 1973.

P.A. Fawcett, J.J. Usher, J.A. Huddleston, R.C. Bleaney, J.J. Nisbet, E.P. Abraham, <u>Biochem. J.</u>, 157, 651-660, 1976.

H.G. Floss in "Antibiotics IV Biosynthesis", J.W. Corcoran Ed. Springer Verlag, Berlin, 1981, pp. 215-235.

H.G. Floss, C. Chang in "Antibiotics IV Biosynthesis", J.W. Corcoran Ed. Springer Verlag, Berlin 1981, pp.193-214.

F. Foor, G.M. Brown, Methods in Enzymology, 66, 303-307, 1980.

Y. Fukagawa, T. Sawa, T. Takeuchi, H. Umezawa, <u>J. Antibiot.</u>, 21, 185-188, 1968.

Y. Fukagawa, T. Sawa, I. Homma, T. Takeuchi, H. Umezawa, <u>J. Antibiot.</u>, 21, 358-360, 1968.

O. Ghisalba, J. Nüesch, <u>J. Antibiotics</u>, 31, 202-214, 1978.

O. Ghisalba, J. Nüesch, <u>J. Antibiot.</u>, 31, 215-225, 1978.

O. Ghisalba, J. Nüesch, J. Antibiotic., 34, 64-71, 1981.

W.O. Godtfredsen, H. Lorck, E.E. van Tamelen, J.D. Willet, R.B. Clayton, Am. Chem. Soc., 90, 208-209, 1968.

S.J. Gould, K.J. Martinkus, C.H. Tann, J. Am. Chem. Soc., 103, 2871-2872, 1981.

S.J. Gould, K.J. Martinkus, C.H. Tann, J. Am. Chem. Soc., 103, 4639-4640, 1981.

S.J. Gould, D.E. Cane, J. Am. Chem. Soc., 104, 343-346, 1982.

M.L. Green, W.H. Elliot, Biochem. J., 92, 537-549, 1964.

J.F. Grove in "Antibiotics II Biosynthesis", D. Gottlieb and P.D. Shaw Eds. Springer Verlag, Berlin 1967, pp. 123-133.

J.M.T. Hamilton-Miller, Becteriol. Rev., 37, 166, 1973.

S.J. Hammond, M.P. Williamson, D.H. Williams, L.D. Boeck, G.G. Marconi, J. Chem. Soc. Chem. Commun. 344-346, 1982.

L.J. Hanka, L.M. Reineke, D.G. Martin, J. Bacteriol., 100, 42-46, 1969.

J.R. Hanson, T. Marten, R. Nyfeler, J. Chem. Soc. Perk. Trans. I, 876-880, 1976.

A. HiroseKumagai, A. Yagita, N. Akamatsu, J. Antibiot., 35, 1571-1577, 1982.

D.J. Hook, R.P. Elander, R.B. Morin in "Peptide anti- biotics, Biosynthesis and functions", H. Kleinkauf, H. von Döhren Eds. de Gruyeter & Co. Berlin, 1982, p. 85-95.

D.J. Hook, L.C. Vining, J. Chem. Soc. Chem., Comm., 185-186, 1973.

V. Hornemann, J.P. Kehrer, J.H. Eggert, J. Chem. Soc. Chem., Commun., 1045-1046, 1974.

L.H. Hurley, M.K. Speedie in "Antibiotics IV Biosynthe- sis", J.W. Corcoran Ed., Springer New York, 1981, pp. 262-294.

L.H. Hurley, C. Gairola, Antimicrob. Agents Chemother. 15, 42-45, 1979.

A. Hurst, J. Gen. Microb. 44, 209-220, 1966.

A. Hurst, G.M. Paterson, Can. J. Microbiol., 17, 1379-1384, 1971).

C.R. Hutchinson in "Antibiotics IV Biosynthesis", J.W. Corcoran Ed. Springer Verlag Berlin, 1981, pp. 1-11.

L. Ingram, Biochim. Biophys. Acta, 224, 263-265, 1970.

K. Isono, S. Funayama, R.J. Suhadolnik, Biochemistry, 14, 2992-2995, 1975.

K. Kakinuma, Y. Ogawa, T. Sasaki, H. Seto, N. Otake, J. Am. Chem. Soc., 103, 5614-5616, 1981.

E. Katz, A.L. Demain, Bacteriol. Rev., 41, 449-474, 1977.

A. Kawaguchi, S. Okuda, J. Chem. Soc. Chem. Commun., 1012-1013, 1970.

J.J. Kibby, I.A. McDonald, R.W. Rickards, J. Chem. Soc. Chem. Commun., 768-769, 1980.

B. Kniep, H. Grisebach, Eur. J. Biochem., 105, 139-144, 1980.

K. Kurahashi in "Antibiotics IV Biosynthesis", J.W. Corcoran Ed. Springer Verlag New York, 1981, pp.325-352.

K. Kurahashi, Ann. Rev. Biochem., 43, 445-459, 1974.

W. Kurylowicz, "Antibiotics. A critical review", Polish Medical Publishers Warsaw, 1976.

G. Lancini, W. Zanichelli in "Structure-activity relation-ships among the semisynthetic antibiotics", D. Perlman Ed. Academic Press New York, 1977, pp. 531-600.

G. Lancini, M. Grandi, "Biosynthesis of Ansamycins" in Antibiotic IV - Biosynthesis, J.M. Corcoran Ed. Springer New York, 1981, pp. 12-40.

G. Lancini, F. Parenti, "Antibiotics. An integrated view", Springer Verlag, New York, 1982.

M.B. Lennon, R.J. Suhadolnik, Biochim. Biophys. Acta, 425, 532-536, 1976.

F. Lipmann, Science, 173, 875-884, 1971.

F. Lepmann, <u>Adv. Microbiol. Physiol</u>., 21, 227-266, 1980.

P.B. Loder, E.P. Abraham, <u>Biochem. J</u>., 123, 477-482, 1971.

F. Lynen, M. Tada, <u>Angew. Chem</u>., 73, 513-519, 1961.

F. Lynen, <u>Biochem. J</u>., 102, 381-400, 1967.

L.A. Mitscher, L.L. Martin, D.R. Feller, J.R. Martin, A.W. Goldstein, <u>J. Chem. Soc. Chem. Commun</u>., 1541-1542, 1971.

J.F. Martin, L.E. McDaniel, <u>Biochim. Biophys. Acta</u>, 411, 186-194, 1975.

J.F. Martin, L.E. McDaniel in "Advances in Applied Microbiology 21", D. Perlman Ed. Academic Press New York, 1977, pp. 2-52.

J.F. Martin, <u>Ann. Rev. Microbiol</u>. 31, 13-38, 1977.

S. Masamune, G. Bates, J.W. Corcoran, <u>Angew. Chem</u>. (Int. Ed. Engl.), 16, 585-607, 1977.

J.R.D. McCornick in "Antibiotics II Biosynthesis", D. Gottlieb and P.D. Shaw Eds. Springer Verlag Berlin, 1967, pp. 113-122.

R. McGrath, L.C. Vining, F. Sala, D.W.J. Westlake, <u>Can. J. Biochem</u>., 46, 587-594, 1968.

A. Meister, S.S. Tate, <u>Ann. Rev. Biochem</u>., 45, 559-604, 1976.

H. Ogawara, K. Maeda, H. Umezawa, <u>Biochemistry</u>, 7, 3296-3302, 1968.

T. Okuda, Y. Ito in "Aminoglicoside Antibiotics", H. Umezawa, I.R. Hooper Eds., pp. 111-203, Springer Verlag Berlin, 1982.

S. Omura, A. Nakagawa, <u>J. Antibiot</u>., 28, 401-432, 1975.

S. Omura, A. Nakagawa, in "Antibiotics IV. Biosynthesis", J.W. Corcoran Ed. Springer Verlag Berlin, 1981, 175-192.

R. Ortmann, U. Matern, M. Grisebach, P. Stadler, V. Sinnwell, H. Paulsen, <u>Eur. J. Biochem</u>., 43, 265-271, 1974.

J. O'Sullivan, A.M. Gillum, C.A. Aklonis, M.L. Souser, R.B. Sykes, Antimicrob. Agents Chemother., 21, 558-564, 1982.

M.L. Souser, R.B. Sykes, Antimicrob. Agents Chemother., 21, 558-564, 1982.

R.C. Paulick, M.L. Casey, H.W. Whitlock, J. Am. Chem. Soc., 98, 3370-3371, 1976.

C.J. Pearce, K.L. Rinehart in "Antibiotics IV Biosynthesis", J.W. Corcoran Ed. Springer Berlin, 1981, pp. 74-100.

D. Perlman, M. Bodanszky, Ann. Rev. Biochem., 40, 449-464, 1971.

D. Perlman, S. Otani, K.L. Perlman, J.E. Walker, J. Antibiot., 26, 289-296, 1973.

G.W.E. Plant, C.M. Smith, W.L. Alworth, Ann. Rev. Biochem. 43, 899-922, 1974.

T.G. Pridham, A. Lyons in "Actinomycetes : the boundary microorganisms", T. Arai Ed., University Park Press, Tokyo, 1976, pp. 373-541.

S.W. Queener, N. Neuss in "The Chemistry and Biology of β-lactam antibiotics, Vol. 3, R.B. Morin, M. Marvin Eds. Academic Press New York, 1982, pp. 1-81.

T. Riisom, H.J. Jakobsen, N. Rastrup Andersen, H. Lorck, Tetrahedron Lett., 26, 2247-2250, 1974.

K.L. Rinehart, R.M. Stroshane, J. Antibiotic., 29, 319-353, 1976.

K.L. Rinehart, M. Potgieter, D.A. Wright, J. Am. Chem. Soc., 104, 2649-2652, 1982.

H. Ristow, J. Salnikow, H. Kleinehauf, FEBS Letters, 42, 127-130, 1974.

L. Sankaran, B.M. Pogell, Antimicrob. Agents Chemother., 8, 721-732, 1975.

J.P. Scannel, D.L. Pruess, T.C. Demny, T. Williams, A. Stempel, J. Antibiotics, 23, 618-619, 1970.

H. Seto, I. Yamaguchi, N. Otake, H. Yonehara, Agric. Biol. Chem., 32, 1292-1298, 1968.

H. Seto, K. Furihata, H. Yonehara, J. Antibiot., 29, 595-596, 1976.

M.R. Stroshane, M. Taniguchi, K.C. Rinehart, I.P. Rolls, W.J. Haak, B.A. Ruff, J. Am. Chem. Soc., 98, 3025-3027, 1976.

T. Sugimori, R.J. Suhadolnik, J. Am. Chem. Soc., 87, 1136-1137, 1965.

R.J. Suhadolnik, "Nucleosides as biological problems", John Wiley and Sons New York, 1979.

R.J. Suhadolnik in "Antibiotics IV" - Biosynthesis, J.W. Corcoran Ed. Springer Verlag New York, 1981.

J.O. Sullivan, E.P. Abraham in "Antibiotic IV - Biosynthesis", J.W. Corcoran Ed. Springer Verlag New York, 1981, pp.

K. Suzukake, T. Fujiyama, H. Hayashi, M. Mori, H. Umezawa, J. Antibiot., 32, 523-530, 1979.

K. Suzukake, M. Mori, H. Hayashi, H. Umezawa in "Peptide Antibiotics - Biosynthesis and functions". H. Kleinhauf, H. von Dören Eds., de Gruyeter & Co., Berlin, 1982, pp. 325-336.

Y. Takeda, H.G. Floss, J. Nat. Prod., 42, 691-692, 1979.

S.W. Tanenbaum in "Antibiotics II - Biosynthesis", D. Gottlieb and P.D. Shaw Eds., Springer Verlag Berlin, 1967, pp. 82 - 112.

C.A. Townsend, A.M. Brown, J. Am. Chem. Soc., 103, 2873-2875, 1981.

W.B. Turner, "Fungal metabolites", Acedemic Press London, 1971.

W.B. Turner in "The filamentous fungi" Vol. I, J.E. Smith, D.R. Berry Eds. Edward Arnold Ltd., 1975, pp. 122-139.

T. Uematsu, R.J. Suhadolnik, Biochemistry, 9, 1260-1266, 1970.

H. Umezawa, I.R. Hooper Eds., "Aminoglicoside Antibiotics" Springer Verlag, Berlin, 1982.

Z. Vanek, J. Majer in "Antibiotics II - Biosynthesis",
D. Gottlieb, P.D. Shaw Eds., Springer Verlag Berlin, 1967,
pp. 154 - 188.

Z. Vanek, Z. Hostaleck in Genetics of Industrial
Microorganisms", Vol. II, Z. Vanek, Z. Hostaleck, J.
Cudlin Eds., Elsevier, Amsterdam 1973, pp. 353-371.

Z. Vanek, J. Tax, J. Cudlin, M. Blumanerova, N.
Steinerova, J. Mateju, I. Komersova and K. Sajner in
Genetics of Industrial Microorganisms, K.D. McDonald Ed.
Academic Press, London, 1976.

Z. Vanek, J. Mateju, J. Cudlin, M. Blumanerova, P.
Sedmera, J. Jizba, E. Kralovcova, J. Tax, G.F. Gauze in
Overproduction of microbial products, V. Krumphanzl,
B. Aikyta, Z. Vanek Eds. Academic Press London, 1982,
pp. 283-299.

H.P. Wahl, U. Matern, H. Grisebach, Biochem. Biophys.
Res. Commun., 64, 1041-1045, 1975.

B. Walker , J. Biol. Chem. 249, 2397-2404, 1974.

J.B. Walker, Methods Enzymol., 43, 429-470, 1975.

D.D. Weller, K.L. Rinehart, J. Am. Chem. Soc., 100,
6757-6760, 1978.

J.W. Westley, Adv. Appl. Microbiol., 22, 177-223, 1977.

J.W. Westley in "Antibiotics IV - Biosynthesis", J.W.
Corcoran Ed. Springer, Berlin 1981, pp. 41-73.

R.J. White, E. Martinelli, G.G. Gallo, G.C. Lancini,
P. Beynon, Nature, London, 243, 273-277, 1973.

R.J. White, E. Martinelli, FEBS Lett., 49, 233-236, 1974.

R.J. White, E. Martinelli; G.C. Lancini, Proc. Natl.
Acad. Sci., U.S.A., 71, 3260-3264, 1974.

P.F. Wiley, D.W. Elrod, V.P. Marshall, J. Org. Chem.,
43, 3457-3461, 1978.

D.F. Witz, E.J. Hessler, T.L. Miller, Biochemistry, 10,
1128-1133, 1971.

E.E. Woodside, J.B.G. Kwapinski in "Molecular Microbiology", Kwapinski Ed., J. Wiley & Sons New York, pp. 129-184, 1974.

R. Zocher, H. Kleinhauf, <u>Biochem. Biophys. Res. Commun</u>., 81, 1162-1167, 1978.

ENZYMATIC PRODUCTION OF SECONDARY METABOLITES

H. Kleinkauf and H. V. Döhren

Institut für Biochemie und Molekulare Biologie
der Technischen Universität Berlin
Franklinstraße 29, D-1000 Berlin 10 (West)

In a recent discussion of the microbiological production of pharmaceuticals, it was pointed out that process improvement in the case of secondary metabolites can hardly be achieved by single gene manipulation, since between 10 and 30 structural genes contribute by the joint action of their products to the biosynthesis of one single complex compound.[1] Reviewing the last 15 years progress on biosynthetic pathways with special reference to gene products or enzymes, we have found many instances of a new type of enzyme organization, i.e. one organized into multienzymes [2] (table 1). A more detailed analysis shows that multienzymes may catalyze up to 8 consecutive reaction steps (table 2).

Table 1. Enzyme organization in the biosynthesis of complex metabolites.

product	structure	reaction steps[1])	number of (multi) enzymes
enniatin	6-depsipeptide	5	1
leupeptin	modified 3-peptide	4	3
bacitracin	modified 12-peptide	18	3
gramicidin S	cyclo-10-peptide	7	2
tyrocidine	cyclo-10-peptide	12	3
patulin	modified polyketide	14	8

1) minimal number of steps required.

107

Table 2. Limiting capacity of multienzymes

synthetase	number of reactions catalyzed minimal[1] actual[2]		amino acid activations	epimeri- zations	peptide/ ester bonds	modifi- cations
gramicidin S - 1	2	3	1	1	—	—
tyrocidine - 1	2	3	1	1	—	—
bacitracin - 2	3	6	2	1	2	—
tyrocidine - 2	4	10	3	1	3	—
enniatin	5	7	2	—	2	1
gramicidin S - 2	6	14	4	—	6	—
tyrocidine - 3	6	18	6	—	6	—
bacitracin - 1	7	16	5	1	4	2
bacitracin - 3	8	18	5	2	6	—

1) minimal number: addition of an amino acid, epimerization, modification.

2) actual number: activation = reactions (adenylate, thiolester), peptide/ ester bond, epimerization, modification.

In this new type of organization, functional units or single gene products are linked to a strict sequence of multifunctional products thus directing a strict sequence of reactions represents an effective assembly line type of pathway to a secondary metabolite. To illustrate the concepts, a few well-characterized examples will be given (Figs. 1 - 4).

The cyclic decapeptide gramicidin S (Fig. 1), still the subject of fundamental biochemical studies, is formed by an enzyme system of Bacillus brevis ATCC 9999. The multienzymes GS 1 (100 KDa) and GS 2 (280 KDa) activate the constituent amino acids at the expense of one ATP-phosphate bond for each peptide bondformed. Following

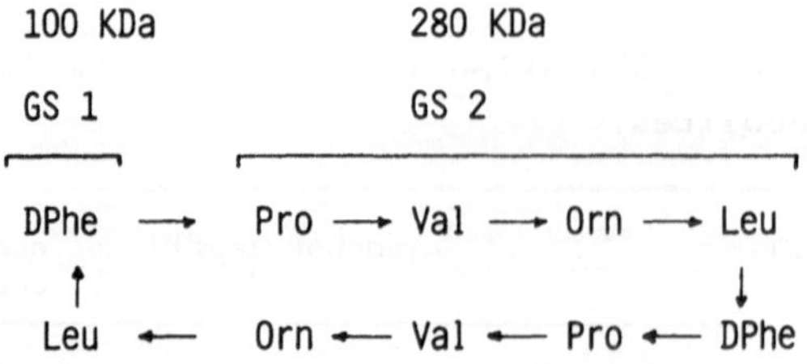

```
100 KDa              280 KDa

GS 1                 GS 2
 ┌─────┐     ┌──────────────────────────────┐
DPhe  ──→   Pro ──→ Val ──→ Orn ──→ Leu
 ↑                                    ↓
Leu  ←──   Orn ←── Val ←── Pro ←── DPhe
```

Fig. 1. Structure and multienzyme organization of gramicidin S. The two multienzymes GS 1 and GS 2 (100 and 280 KDa) catalyze cyclopeptide formation from the constituent L-amino acids and ATP.

epimerization of the starter amino acid, phenylalanine, initiation of chain growth proceeds by formation of the enzyme-bound activated dipeptide phenylalanyl-proline. Transport of this and the following intermediate peptides is facilitated by the enzyme-attached cofactor 4'-phosphopantetheine. The symmetrical decapeptide is formed from pentapeptides by head-to-tail cyclization.

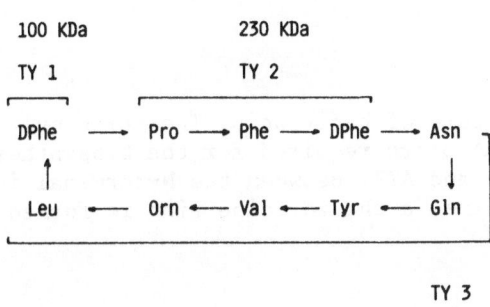

Fig. 2. Structure of the cyclodecapeptide tyrocidine and the organization of the corresponding multienzymes TY 1, TY 2 and TY 3 with sizes of 100, 230 and 450 KDa.

Tyrocidine (Fig. 2) is the cyclic constituent of tyrothricin, which is produced by Bacillus brevis ATCC 8185 and is used as a topical antibiotic. The enzyme system consists of 3 multienzymes activating 1, 3 and 6 amino acids, respectively. The termination reaction is the cyclization of the enzyme-bound activated decapeptide. The sequence homologies to gramicidin S indicate corresponding homologies within the multienzymes.

Bacitracin (Fig. 3), a 7-cyclo dodecapeptide from strains of Bacillus licheniformis or B. subtilis, is in use as a topical antibacterial and animal feed supplement. Its biosynthesis is accomplished by a set of 3 multienzymes with sizes of 330, 210, and 380 KDa activating 5, 2, and 5 amino acids, respectively.

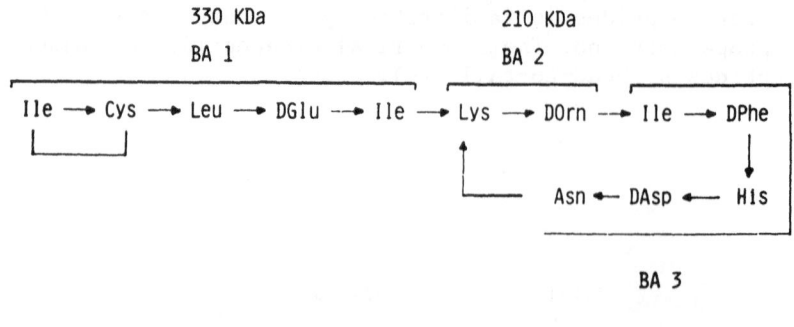

Fig. 3. Structure of bacitracin. The three multienzymes BA 1, BA 2, and BA 3 are required for the biosynthesis from L-amino acids and ATP. Between the N-terminal isoleucine and cysteine, a thiazolidine ring is formed.

Leupeptin (Fig. 4) is a modified tripeptide with proteinase inhibiting properties produced by Streptomyces roseus. Contrary to the other examples, its biosynthesis does not involve covalent intermediates. An acyl-transferase, a tripeptide synthetase of 260 KDaltons, and a reductase have been characterized.

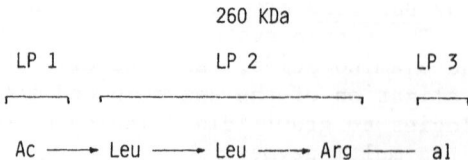

Fig. 4. Leupeptin structure and its biosynthesis. A leucyl-acetyl-transferase (1) produces the starter substrate from acetyl-CoA which is elongated on a multienzyme (2, 260 KDa) and finally reduced to the aldehyde (3).

110

Comparing the chemistry of peptide formation in enzymic and nonenzymic systems, we recollect that a number of unit operations are required to make one peptide bond. Thus activation of a carboxyl group and protection of one amino group are followed by peptidation and deprotection. In the enzymic system, activation is achieved by enzyme attachment and protection is afforded by noncovalent interactions with the protein(Fig.5).

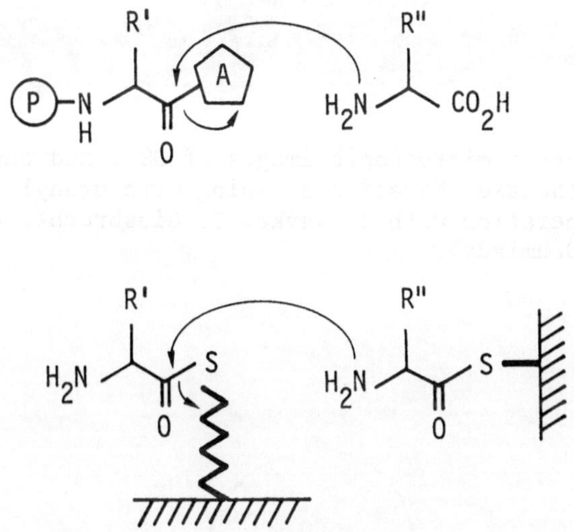

Fig. 5.   Comparing chemical (top) and enzymatic peptide formation ((P): Protection A : Activation); Note the 2 different types of thiolesters in the enzymic peptidation reaction.

The sequence of reactions is determined by the structural organization of the multienzymes involved. Electron microscopic studies of gramicidin S-synthetase 2 and enniatin synthetase revealed compact globular structures with ring-like appearances (Fig. 6/7)[3].

Present information gives the following working hypothesis for the structure of GS 2 (Fig. 8). The amino acid activating functional domains are arranged around the pantetheine-containing carrier.

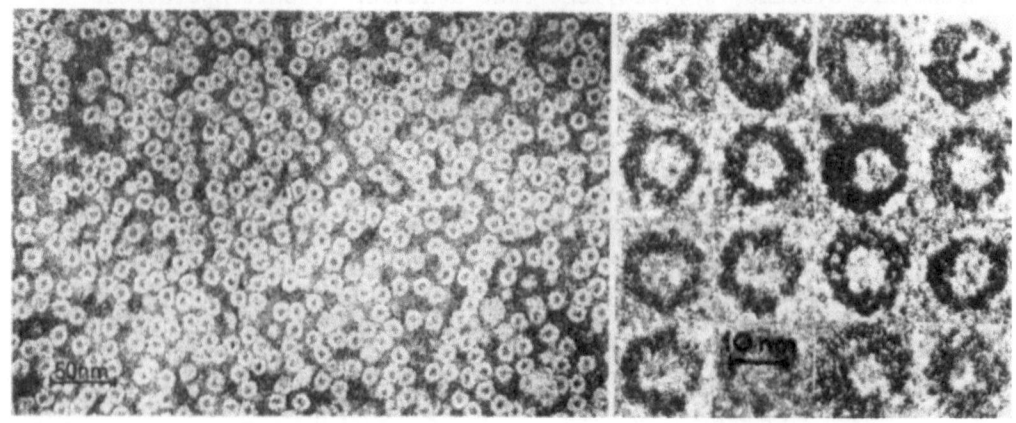

Fig. 6/7.   Electron microscopic images of GS 2 and enniatin
synthetase. Negative staining with uranyl acetate (in
cooperation with J. Wevke, P. Giesbrecht, B. Tesche and
H. Schmiady).

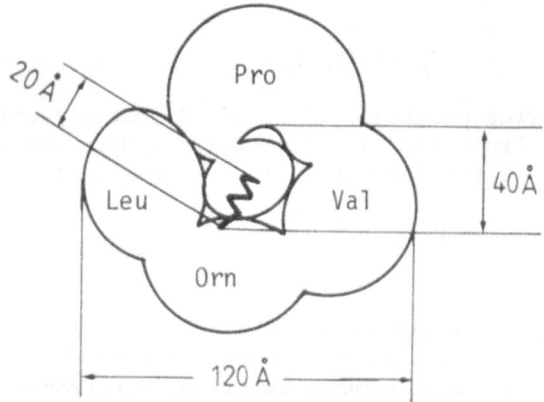

Fig. 8.   Structural model of the multienzyme GS 2. Predicted
arrangement of functional units in relation to electron
microscopic images. All active centers have to be in
reach of the cofactor (length 20 Å).

112

Structural analysis is carried out with the use of mutationally altered multienzymes (table 3) and fragmentation by various proteinases (table 4). Mutations lead to multienzymes unchanged in size but with altered properties, multienzyme fragments in early termination mutants, and structures with the cofactor pantetheine missing. Treatment of multienzymes in the native state by various proteinases produces fragments that may contain partial activities. Large fragments containing several activities by either procedure can be used to deduce the arrangement of functional units on the structure. Functional studies are carried out to characterize the mechanism of carboxyl activation and the properties of the resulting thiolesters. The activation proceeds by cleavage of the. ᅄ-B phosphate bond of ATP with formation of highly reactive aminoacyladenylates as intermediates. The adenylates are then stabilized as thiolesters on the enzyme surface. These reactive intermediates may undergo side reactions with certain nucleophiles, e.g. in intramolecular cyclizations. Side reactions that may become significant are cyclization reactions involving the unprotected amino groups. Thus thiol-activated ornithine may be lost from the multienzyme by a ring-size favoured cyclization that is not observed if lysine has been used to replace ornithine (Fig. 9).

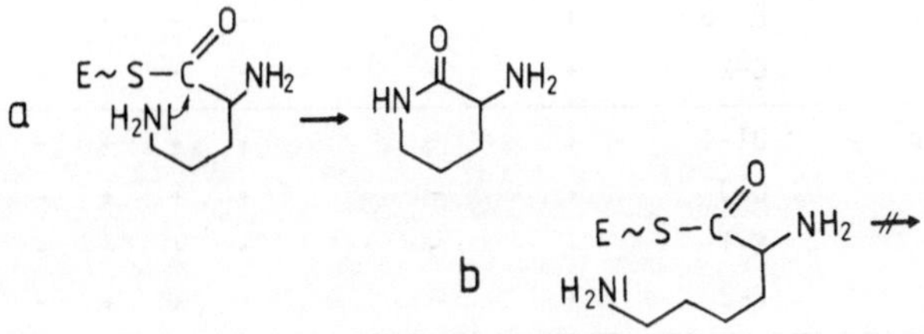

Fig. 9.   Cyclization of activated ornithine is favoured by 6-ring formation (a). Thiol-ester bound lysine on the other hand is a stable intermediate (b).

From studies of mutationally-altered multienzymes carried out by SAITO's group in Japan, altered stabilities of activated intermediates have been detected.[4]

Another side-reaction, well-known from solid phase peptide synthesis, is dipeptide cyclization, especially of proline-containing peptides. The initiating peptide -D-Phe-Pro in gramicidin S and tyrocidine biosynthesis can be lost as diketopiperazine (Fig. 10). Rates of this side reaction, as compared to formation of the decapeptides, can be enhanced if certain analogs of proline are used that favour cyclization (Fig. 11).

Table 3. Gramicidin S-Nonproducer Mutants

Group			Strain	GS 1 (100 KDa)		GS 2 (280 KDa)					
				1) Phe	2) L-D	1) Pro	1) Val	1) Orn	1) Leu	3) Pan	4) Size
I	GS 1⁻	GS 2⁻	BI-1	-		-	-	-	-		
			BII-4	-		-	-	-	-		
			C 4	-		-	-	-	-		
			aa	-		-	-	-	-		
II	GS 1⁻		BI-7	-		+	+	+	+		
			C-I	-		+	+	+	+		
			E-3	-		+	+	+	+		
			a-10	-		+	+	+	+		
III	GS 2⁻		BI-5	+	+	-	-	-	-		
			C-2	+	+	-	-	-	-		
IV	GS 1	Pan⁻	BI-4	+	-	+	+	+	-		
			C-3	+	-	+	+	+	-		
			E-1	+	-	+	+	+	-		
			E-2	+	-	+	+	+	-		
VI	GS 2		BII-3	+	+	-	+	+	+	+	280
			BI -3	+	+	+	-	+	+	+	280
			BI -6	+	+	+	-	+	+	+	
			BII-1	+	+	+	-	+	+	+	
			E -4	+	+	+	-	+	+	+	
			E -5	+	+	+	-	+	+	+	
			BI -9	+	+	+	+	+	-	+	280
			BI -2	+	+	+	+	+	+		
			n⁷	+	+	+	-	-	-		100
			hh	+	+	+	+	+	-		260

1) activation of the amino acid indicated
2) epimerization of the activated amino acid
3) cofactor 4'-phosphopantetheine
4) size as measured by SDS-polyacrylamide gel electrophoresis

Table 4. Effects of Various Proteinases on the Catalytic Activities of the Multienzyme GS 2

	Biosynthesis with complementary GS1[1] (cpm × 10^{-3})	ATP-PP$_i$ exchange reaction dependent on (cpm × 10^{-3})[2]			
		Pro	Val	Orn	Leu
0 control	10.4	25.4	32.0	16.9	34.5
1 leucine aminopeptidase	–[3]	22.2	36.1	16.5	38.8
2 pepsinogen	1.1	31.9	40.9	15.4	30.1
3 pancreatic crude	0.5	21.5	18.0	5.6	21.4
4 *Streptomyces caespitosus* proteinase	11.1	31.5	37.7	16.9	36.1
5 *Streptomyces griseus* proteinase	1.3	20.5	38.3	12.9	33.1
6 Subtilisin carlsberg	0.2	20.3	25.8	5.2	19.8
7 *Bacillus polymyxa* proteinase	6.5	34.3	40.5	14.1	40.0
8 Thermolysin	12.0	25.9	37.4	14.2	32.2
9 *Tritirachium album* proteinase	0.2	10.3	22.7	3.4	13.1
10 papain	0.2	23.7	29.5	6.6	31.0
11 bromelain	0.3	10.2	12.5	9.6	5.2
12 elastase	12.7	29.1	42.4	16.7	38.6
13 chymotrypsin	1.2	29.1	6.4	10.7	13.7
14 trypsin	0.5	–	–	–	–

1 reaction conditions were GS2: proteinase 100: 1,30 minutes at 27°C preincubation, addition of GS1 and substrates, 10' at 37° C INCUBATION,

2 reaction conditions were as in 1 except that the reaction was slowed down by addition of 2mg/ml casein before the assay was performed.

3 not determined

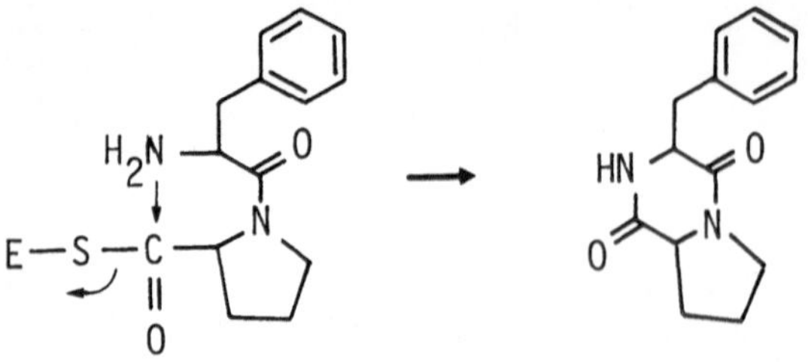

Fig. 10. Cyclization of the proline-containing initiating dipeptide in gramicidin S and tyrocidine biosynthesis.

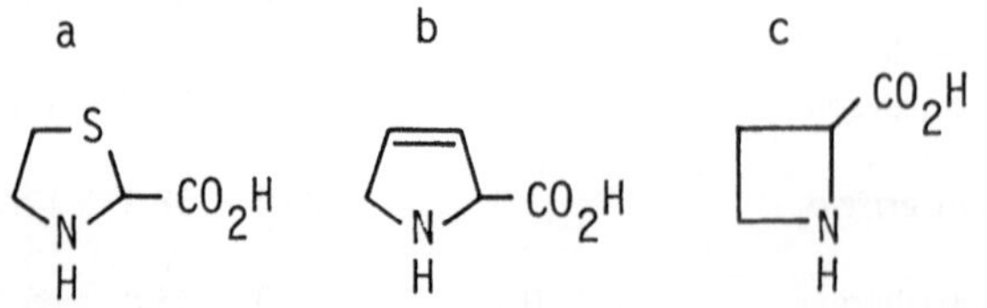

Fig. 11. Analogs of proline that favour dipeptide cyclization of DPhe-X to decapeptide formation by a factor of 8 (a, B-thioproline), 4 (b, 3,4-dehydroproline) and 3 (c, azetidine-2-carboxylic acid).

     As will be discussed later, enzymatic synthesis of analog compounds should be optimized in each particular case. The broad specifity permits many substitutions that may be useful for solubility properties or modification reactions.

     The limiting factor of synthesis of peptide analogs using an isolated enzyme system is substrate binding and processing. Several substrates may be activated but are not incorporated into peptides. Possible controls are thiolester formation, initiation (amino acid transfer), and peptide transfer. The molecular mechanism of sequential peptidation reaction is still obscure. From the structural comparison of multienzymes like gramicidin S-synthetase and tyrocidine synthetase, partial identities of the polypeptide chains can be expected, regarding the location of the transport function 4'-phosphopantetheine and amino acid activation sites (Fig. 12). Thus a more general type of control of sequential reactions is expected rather than a highly specialized system adapted to one particular product. In 1973, Zharikova et al. reported a spontaneous mutant from the gramicidin S producer B. brevis that formed gratisin [5] (Fig. 13).

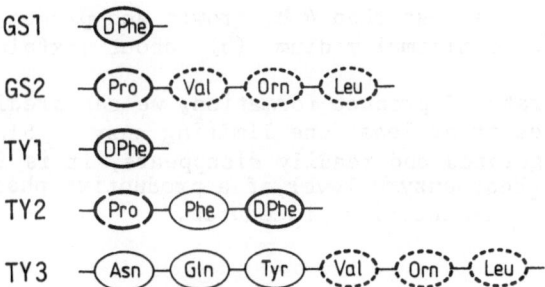

Fig. 12.   Expected partial identities of amino acid activating sites on multifunctional peptide synthetases involved in gramicidin S and tyrocidine biosynthesis.

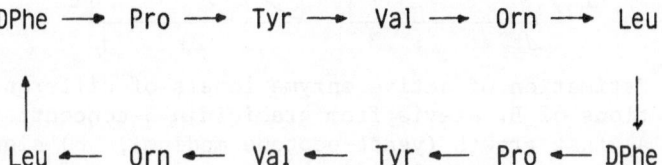

Fig. 13.   Gratisin. The D-configurations of Phe and Tyr have been concluded from the synthetic studies of Tamaki et al.[11] [12]

At present we are far from understanding the molecular mechanisms leading to possible alterations of multienzyme structures. Sequence homologies, however, are frequently observed comparing microbial peptides. More detailed studies are needed to understand the functioning of enzyme systems, their structural organization, and their possible manipulation.

PRODUCTION OF MULTIENZYME SYSTEMS

The isolation of a secondary metabolite's enzyme system requires the preparation of enriched cell material. As a first orientation, conditions of high productivity can be chosen. Comparing fermentation patterns, production of gramicidin S concentrations of several g/l can be achieved in different times depending on growth conditions (Fig. 14). While in a yeast extract/peptone complete medium (a) growth is completed within 10 h of fermentation, and the actual production phase is less than 4 h, growth and production are slowed down in a fructose minimal medium (b) about sixfold.

From the rate of product formation, we can predict the actual level of enzymes or at least one limiting enzyme. Since enzyme activity is regulated and readily disappears, it is important to predict the highest enzyme level of a productive phase or to stabilize such a phase by fermentation procedures.

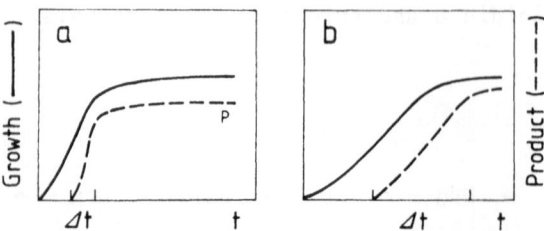

Fig. 14.   Estimation of active enzyme levels of different fermenta-
           tions of B. brevis from gramicidin S-concentration (--);
           a) fast growth (yeast-peptone medium), b) slow growth
           (fructose-minimal medium).

A detailed study, however, of actual enzyme concentrations is quite important. Consider as an example, the activity of bacitracin synthetase of Bacillus licheniformis in a glutamine minimal medium (Fig. 15).

The highest activity of the soluble enzyme system does not coincide with the bacitracin production phase. Of some help in monitoring the presence of the enzyme systems is their unusual size which can easily be detected by polyacrylamide gel electrophoresis using denaturating conditions (Fig. 16).

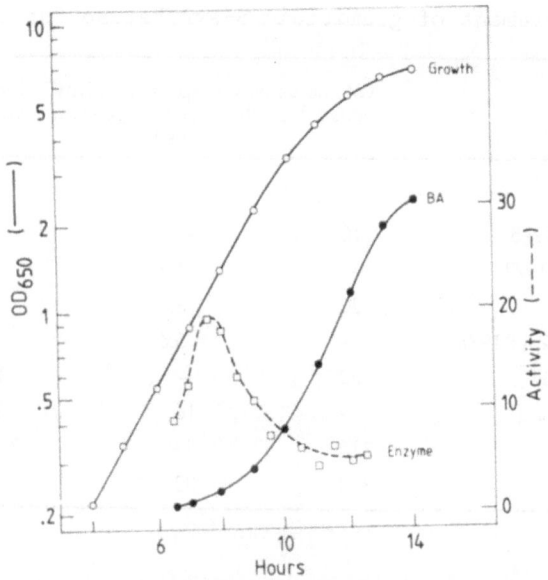

Fig. 15. Bacitracin fermentation. Correlation of enzyme activity
and product formation [6] .

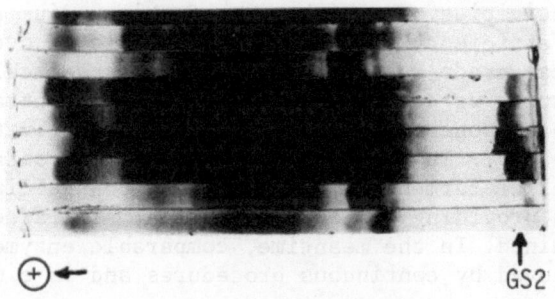

Fig. 16. SDS-PAGE of GS 2

        Much effort has been conducted to improve the fermentation of
B. brevis for the production of gramicidin S-synthetase. At M.I.T.[7],[8]
and later at the TU Berlin [2],[9],[10] , batch and continuous procedures
now give about 100 fold higher activities than during the initiating
studies in the sixties (table 5).

Table 5. Improvement of gramicidin S-synthetase fermentations.

	volume activity units / 1 broth	specific activity units / g cells (wet)	fermentation volume
Otani et al. 1970	—	0.035	?
Zimmer & Laland 1975	10	—	10
Christiansen et al. 1977	—	1.3	10
Tomino et al. 1967	20	3.7	10
Koischwitz & Kleinkauf 1976	—	5.6	1
Bredesen et al. 1968	50	—	120
Tzeng et al. 1975	—	16	180
Aust 1982	710	32	3.000
Chiu 1983	300	32	10 (continuous)

Frequently two types of media have been employed in gramicidin S-synthetase fermentations: a minimal medium based on glutamine, and a rich medium with yeast extract/peptone introduced by the M.I.T.-group in 1975 [7]. The latter medium consists of 5 % yeast extract and peptone and yields up to 30 g of wet cells/1 medium with high enzyme activities. The high cost of this medium and the extremely low efficiency in substrate use could be overcome by the use of technical grade substrates. During a large scale fermentation of 3,000 1 volume, we cut down the substrate concentrations to 1.25 % of each component; and excellent enzyme activities were still recovered (table 6). It proved quite difficult to obtain similar results in smaller fermenters. Recently, by controlling the oxygen level and thus providing a slow growth rate, similar enzyme levels have been obtained. In the meantime, comparable enzyme levels have also been achieved by continuous procedures and in a minimal medium.

In the minimal medium, a discrete optimum of phosphate concentration for gramicidin S-production has been described [10]. As a general observation, phosphate exerts some type of control on the production of secondary metabolites (Fig. 17). We have been able to directly correlate this control with enzyme concentration. To our surprise, we found a second high enzyme production peak at a high phosphate concentration (Fig. 18), well beyond all known phosphate optima. This effect was not observed when NaCl was substituted for sodium phosphate. Further fermentation studies indicate that a slow growth

Table 6. Production of gramicidin S-synthetase on yeast/peptone media.

medium concentration	fermenter volume (1)	DCW (g/1)	specific activity units/g cells (DCW)	units/mg protein (crude extract)
2.5%	3.000	5.6	129	0.46
	500	4.2	39	0.10
10 %	80	3.8	126	0.53
2.5%	8	4.5	46	0.26
2.5%[x]	8	4.0	38	0.34
10%[xx]	12	6.0	83	0.34

[x] technical grade yeast extract

[xx] B. brevis (Oslo)

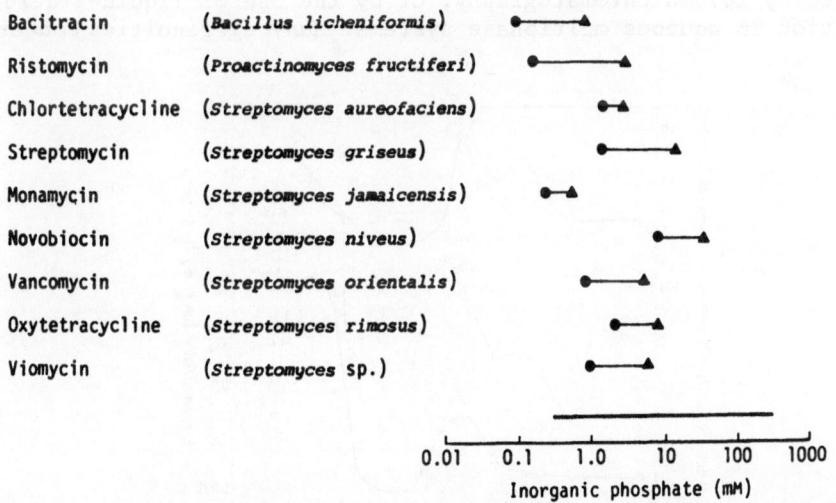

Fig. 17. Phosphate control of the production of various antibiotics. Maximal production at this or lower concentration (●); no production (▲). The solid line at the bottom indicates optimal growth concentrations of both procaryotes and eucaryotes [13].

phase, that can be achieved by high phosphate (buffer capacity), pH-control or oxygen limitation, can be used for a significant increase of multienzyme levels. With the isolated enzyme system, higher product yields can be achieved than occurs in vivo.

Parameters that have been used to control the state of the culture and to estimate the proper time of harvest are growth rate, pH, dissolved oxygen tension, and redox potential. The continuous procedure limits the cell concentration by a limiting substrate.

Thus specifically restricted growth conditions can be established in steady state. It has been found that regardless of the type of limitation used, gramicidin S-synthetase could be induced [7]. It could be verified that enzyme formation and sporulation are unrelated processes (Fig. 19), as had also been concluded from the study of mutants not producing the peptide but sporulating normally.

In a medium using fumarate as carbon source introduction of 1 % glycerol and optimized oxygen limitation (table 7, Fig. 20) improved the specific enzyme activities comparable to batch procedures [10].

The isolation of gramicidin S-synthetase, a labile intracellular enzyme system, has been achieved either by conventional procedures using ammonium sulfate precipitation, centrifugation techniques followed by column chromatography, or by the use of liquid-liquid extraction in aqueous multiphase systems. Many difficulties concerning

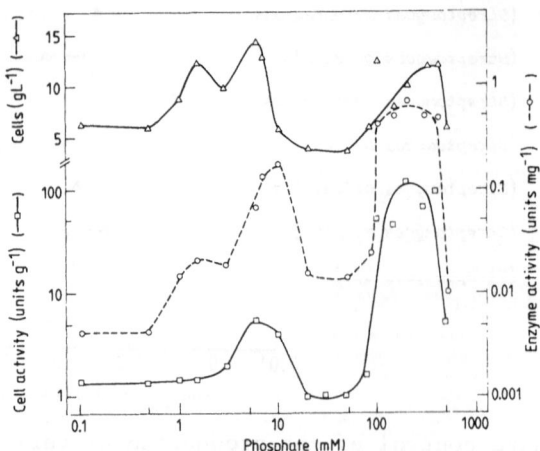

Fig. 18.   Gramicidin S-synthetase activity in glutamine minimal medium dependent on phosphate concentration.

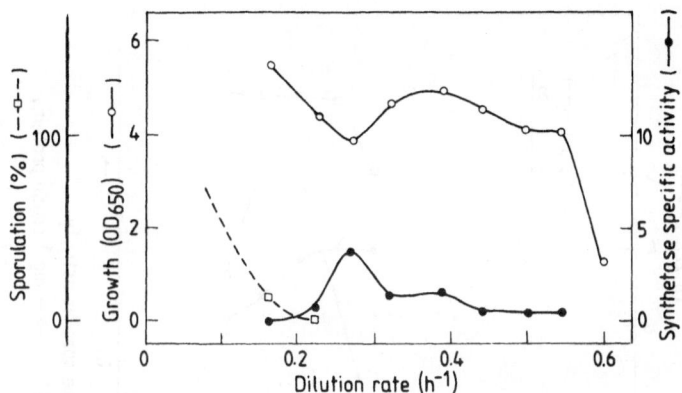

Fig. 19. Continuous fermentation of B. brevis. Limitation of carbon
          source (fumarate). Sporulation is observed under
          conditions where no synthetase is formed.

Table 7. Effect of carbon source on gramicidin S-synthetase
          production.

C-sources	growth OD 620	pH (end)	time (h)	units/g DCW
fumarate	0.50	8.0	12	0.1
succinate	0.48	7.8	9	0.1
glycerol	poor			
glycerol (1%) + fumarate (0.1%)	0.56	6.8	25	1.07
glycerol (1%) + fumarate (0.2%)	0.61	7.6	12	5.73
glycerol (1%) + succinate (0.1%)	0.62	6.65	12	6.75
glycerol (1%) + succinate (0.2%)	0.63	6.80	10	5.60

reproducibility of extraction steps with different batches of cell
material have been observed. An obvious difficulty is that PEG/
dextran-extraction systems use high salt concentrations that
completely inhibit peptide synthesis. In optimizing the salt
dependence and relation to the step following extraction, a 3-phase
system has been selected, using Ficoll. No salt is used and the
enzyme system is recovered in good yield in the Ficoll phase. The

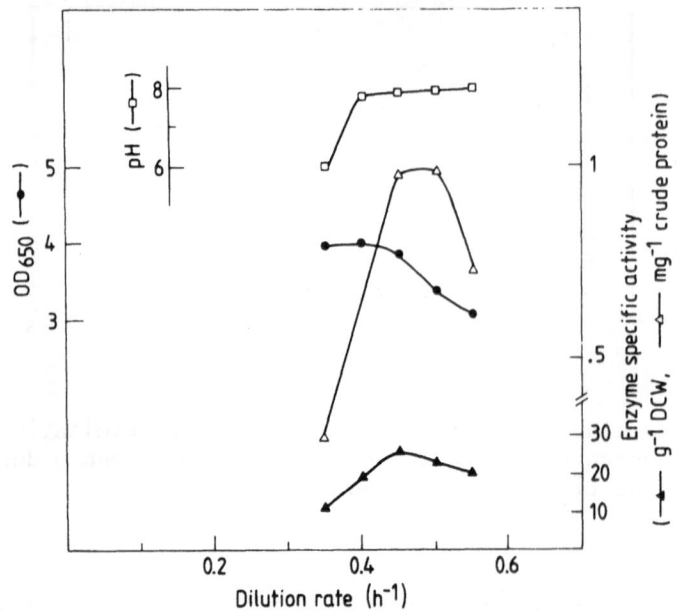

Fig. 20.   Gramicidin S-synthetase formation. Optimization of the
continuous fermentation at a selected oxygen supply.

high costs of dextran and Ficoll may be cut down by the use of
technical dextran and recycling. A comparison of extraction procedures
is given in table 8. Stability of the synthetase in the Ficoll-phase
is sufficient to permit work at room temperature.

Enzymatic Production of Peptides

The considerable improvement of enzyme production achieved
awaits similar progress in the enzymic production of various
products. Most of the available amino acids have been checked as
substrates for gramicidin S-synthetase. Of about 60 amino acids
and analogs. 34 have been incorporated into gramicidin S-like
peptides. These theoretically permit in different combinations,
synthesis of more than 40,000 homologous structures. Not all peptide
structures have been proved so far to be gramicidin S-homologue
structures. With a crude enzyme system the reaction proceeds

Table 8.   Comparison of 2 Extraction Procedures for Gramicidin
           S-synthetase

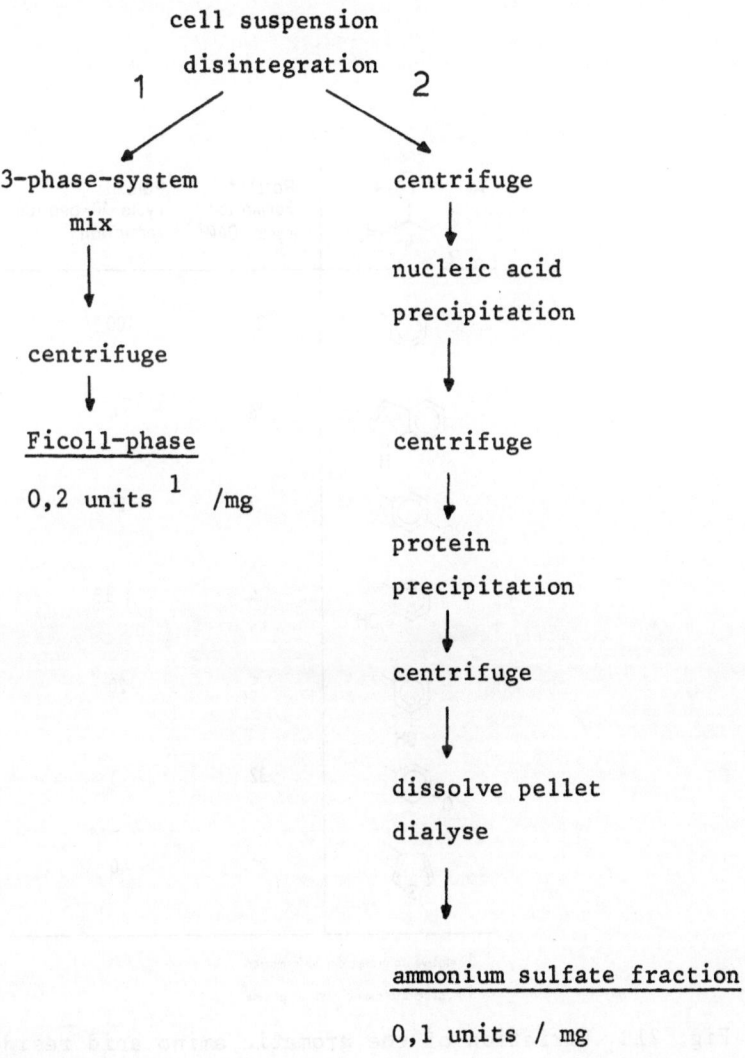

cell suspension

disintegration

1   2

3-phase-system   centrifuge

    mix

                                  nucleic acid
                                  precipitation

    centrifuge

    Ficoll-phase   centrifuge

    0,2 units [1] /mg

                                  protein
                                  precipitation

                                  centrifuge

                                  dissolve pellet
                                  dialyse

                                  ammonium sulfate fraction

                                  0,1 units / mg

---

[1] unit : 1 nM product x min^{-1}

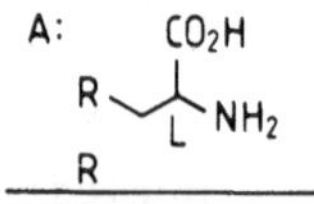 A: R—CH(NH$_2$)—CO$_2$H (R,L)	Rate[1] of Formation cyclo-DAP[2]	Rate of cyclo-10-peptide formation
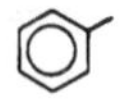 (phenyl)	20	100
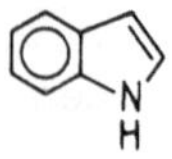 (indolyl, N–H)	16	4
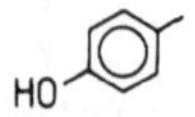 (HO– phenyl, para)	7	29
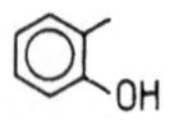 (phenyl, ortho-OH)	4	3.5
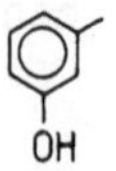 (phenyl, meta-OH)	5	20
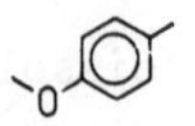 (methoxy phenyl)	32	1
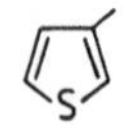 (thienyl, S)	21	40

[1] relative to gramicidin S-formation

[2] A: aromatic amino acid P: proline

Fig. 21. Variation of the aromatic amino acid residue in formation of 2,5-piperazine-diones and gramicidin S-like peptides.

according to the following:

2 Phe

2 Pro

$$2 \ \text{Val} + 10 \ \text{Mg ATP}^{2-} \longrightarrow \text{GS} + 10 \ \text{AMP} + 10 \ \text{Mg pp}_i^{2-}$$

2 Orn

2 Leu

Linear reaction rates can be observed for several hours. The presence of substrates, stabilizes the enzyme system by about a factor of 10. In the crude system, ATP utilization of up to 70 % has been achieved. A partially purified system revealed operation times of up to 30 hours at 37°. However, the instability of enzymes does not yet permit continuous production.

Enniatin synthetase of Fusarim oxysporum has been used for the production of various enniatins according to the following:

3 R

$$3 \ \text{DHiv} + 6 \ \text{Mg ATP}^{2-} \longrightarrow \text{enniatin} \ (\text{R}) + 6 \ \text{AMP} + 6 \ \text{Mg pp}_i^{2-} +$$

3 SAM

3-S-adenosylhomocysteine

(R : Leu, Val, Ile)

Partial stabilization of this single multienzyme has been achieved by adsorption onto propyl-sepharose.

Generally substitution of an amino acid by an analog slows down the reaction rate as shown in Fig. 21. The combined introduction of 2 or more analogs leads to even lower production rates. The rates can roughly be estimated by the product of the individual rates (table 9).

Although there are possible applications, the low stability of multienzymes as well as the inactivation of interacting multi-enzymes by immobilization imposes severe limits. Smaller amounts of peptides or analog peptide structures can be prepared quite easily. These conclusions are based, however, mainly on work with the gramicidin S-system. Properties of other pathways producing metabo-lites or key intermediates with this highly efficient type of system remain to be established.

REFERENCES

1. Hopwood, D.A., and Chater, K.T. (1980) Phil. Trans. Roy. Soc. London B 290, 313
2. Kleinkauf, H., and Döhren, H. von (Eds.) (1982) Peptide Antibiotics-Biosynthesis and Functions. De Gruyter Berlin
3. Kleinkauf, H., Döhren, H. von, Zocher, R., and Tesche, B. (1981)
4. Hori, K., Kanda, M., Kurotsu, T., Miura, S., Yamada, Y., and Saito, Y. (1981) J. Biochem. 90, 439
5. Zharikova, G.G., Myaskovskaya, S.P., and Silaev, A.B. (1972) Vestn. Mosk. Univ. Biol. Puchvoved 27, 110
6. Kleinkauf, H., and Döhren, H. von (1983) in Biochemistry and Genetic Regulation of Commercially Important Antibiotics (L.C. Vining, ed.) pp. 95-145, Addison-Wesley, Reading
7. Tzeng, C.H., Thrasher, K.D., Montgomery, J.P., Hamilton, B.K. and Wang, D.I.C. (1975) Biotechnol. Bioeng. 17, 143
8. Matteo, C.C., Cooney, C.L., and Demain, A.L. (1976) J. Gen Microbiol. 96, 415
9. Aust, H.J., and Döhren, H. von (1982) in Ref. 2. p. 137
10. Chiu, C.W., Bernhard, T., and Dellweg, G. (1982) in Ref. 2, p. 149
11. Tamaki, M., Takimoto, M., Sofuku, S., and Muramatsu, I. (1981) J. Antibiot. 34, 1227
12. Tamaki, M., Takimoto, M., Sofuku, S., and Muramatsu, I. (1983) J. Antibiot. 36, 751
13. Iwai, Y., and Omura, S. (1982) J. Antibiot. 35, 123

NONANTIBIOTIC APPLICATIONS OF MICROBIAL SECONDARY METABOLITES

Arnold L. Demain

Fermentation Microbiology Laboratory
Department of Nutrition and Food Science
Massachusetts Institute of Technology
Cambridge, MA 02139

INTRODUCTION

For years the predominant application of microbial secondary metabolites was that of antibacterial, antifungal and antitumor chemotherapy. However, antibiotic activity represents a small part of the potential application of microbial secondary metabolites for the benefit of humankind. In this article, I shall point out applications other than as antibacterial, antifungal and antitumor agents. These include use against parasites, insects and weeds, as well as for animal and plant growth stimulation, immunosuppression, uterocontraction and many other pharmacological activities. The future will see further applications in various areas of pharmacology and agriculture, a development which is being catalyzed by the use of simple enzyme assays as prescreens prior to testing in intact animals or in the field.

Antiparasitic Activities

One of the major economic diseases of poultry is that of coccidiosis caused by species of the parasitic protozoan <u>Eimeria</u>. Although synthetic chemicals were effective against coccidia, resistance developed rapidly and new chemical modifications of the existing coccidiostats had to be made continually. Surprisingly, a parenterally toxic and narrow spectrum antibiotic, monensin, was found to have extreme potency against coccidia.[1] As a result, the polyethers,[2] especially monensin (produced by <u>Streptomyces cinnamononsis</u>) and lasalocid (produced by <u>Streptomyces lasaliensis</u>), dominate the commercial coccidiostat field today.

An interesting sidelight of the monensin story is the discovery of its further use as a growth promotant in ruminants. Its
use in cattle and sheep diets eliminates wasteful methane production and increases volatile fatty acid formation (especially
propionate) in the rumen, thus improving feed efficiency.[3]

Another major agricultural problem has been the infection of
farm animals by worms. Although certain antibiotics possess
antihelmintic activity (e.g., hygromycin, myxin, thaimycin,
axenomycin) against nematodes or cestodes,[4] these failed to
compete with the synthetic compounds. A recently discovered
organism, Streptomyces avermitilis yielded a broth which killed
the intestinal nematode, Nematosporoides dubius, in mice and was
nontoxic and without antibiotic activity against bacteria or
fungi.[4] The broth contained a family of secondary metabolites
named "avermectins," which are macrocyclic lactones with excep-
tional activity against parasites, i.e., at least ten times higher
than any synthetic antihelmintic agent known. Avermectins appear
to interfere with neurotransmission in many invertebrates.[5] They
have activity against both nematode and arthropod parasites in
sheep, cattle, dogs, horses and swine. A semi-synthetic derivative, 22,23-dihydroavermectin B_1 ("ivermectin")[6] is one thousand
times more active than thiobenzole and is already being used in
certain countries. These potent molecules have an additional
activity as insecticides and may be useful in protecting plants.[7]

Insecticides and Herbicides

The selective toxicity of the crystal protein (i.e., the
delta-endotoxin) of Bacillus thuringiensis against insects of the
order Lepidoptera has been exploited commercially for several
years.[8] This insecticidal toxin has not disturbed the environment
and no resistance has developed. Other potential applications
include the activity of certain strains of Bacillus thuringiensis[8]
and Bacillus sphaericus against mosquitos[9] and Bacillus popillae
against the Japanese beetle.[10]

With regard to low molecular weight microbial metabolites,
there is considerable interest in the potential use of nikkomycin
against agricultural pests. The nikkomycins are nucleoside analogues, structurally related to the polyoxins which are being
used as agricultural antifungal agents. Nikkomycins inhibit
chitin synthetase and thus have potent insecticidal activity,[11]
since chitin is an important structural material for insects.[12]
Other fermentation products with insecticidal activity include
the prasinons[13] and the milbemycins.[14] The macrotetralide, tetra-
nactin, has been in use since 1974 as a mitocide for plants.[15]

Streptomycete secondary metabolites that are herbicidal
include the herbimycins (ansamycins active against mono- and
dicotyledonous plants[16]) and the herbicidins (nucleoside analogues active against dicotyledons[17]).

Plant Growth Regulators

Gibberellins are a group of phytotoxic mycotoxins, produced
by Gibberella fujikuroi, which is the cause of the "foolish
seedling" disease of rice.[18] The gibberellins have been successfully exploited in regulating the growth of plants. They are
used to reduce the time needed for malting of barley, to improve
the quality of malt, and to increase the yield of vegetables as
well as allow their earlier marketing. Development of biennials
can be accelerated to produce seed crops (e.g., lettuce and sugar
beets) in one year instead of two.

Pharmacological Action of Microbial Metabolites

There has been a reluctance to exploit the pharmacological
activities of fermentation broths for the following reasons:
(a) these activities are normally assayed in living animals and
pharmacologists are reluctant to administer crude broths to animals. Pharmacologists prefer screening synthetic chemicals since
there are fewer side-effects and if activity is observed, they
immediately know the structure of the active agent; (b) there is
a bias that microbial metabolites are only useful in solving
microbial problems. However, certain microbial metabolites are
indeed useful in medicine. Their activities were detected
because the products had antibacterial or antifungal activities
although not suitable for use as antibiotics. Since the products
had been purified during the attempt to develop them as antibiotics, there was no reluctance to test such purified materials
for pharmacological activities in animals. As a result,
cyclosporin A (an antifungal antibiotic produced by Tolypocladium
inflatum), is used today as an immunosuppressive agent in human
organ transplants.[19]

The ergot alkaloids, a group of mycotoxins responsible for
poisoning of people eating bread from contaminated grain or animals feeding on contaminated grain or infected grass throughout
the ages, are now used for uterocontractant activity in obstetrics,
to treat migraine headaches, hypertension, serotonin-related
disturbances, to inhibit prolactin release in treating agalactor-
rhoea, and to prevent implantation in early pregnancy.[20] They
are produced by various species of Claviceps in large-scale industrial fermentations. Alkaloids produced by actinomycetes[21]
possess antihistaminic, hypotensive and hypoglycemic activities.

Another mycotoxin whose potent activities have been harnessed is zearelanone, produced by Gibberella zeae.[22] This compound is an estrogen and is used as an anabolic agent in cattle and sheep, improving both growth and feed efficiency.

Microbes are also known to produce broths possessing the activities of neurotensin, human chronic gonadotrophin, insulin, ACTH, beta-endorphin, somatostatin, relaxin and insect juvenile hormone.

Anti-inflammatory activity. A number of actinomycete and Bacillus products have anti-inflammatory activity as measured by inhibition of rat foot pad edema induced by carrageenin.[23-26] One of these is amicomacin A, a Bacillus pumilis antibiotic which shows both anti-inflammatory and anti-ulcer activities[25]; others are forphenicine and esterastin.[26] Other compounds found to show anti-inflammatory activity are the pyrrothin antibiotics produced by Streptoverticillium sp.[27]

Hypocholesteremic activity. The ability to inhibit cholesterol formation in the liver of rats has been observed with citrinin, a metabolite of Pythium ultimum and with compactin, an antifungal agent produced by Penicillium brevicompactum and Penicillium citrinum. Compactin has low acute toxicity and shows activity also in hens and dogs.[28] More recently discovered metabolites include monacolin K from Monascus ruber, a nontoxic metabolite structurally similar to but 4-5 times more active than compactin.[29] Mevinolin, discovered independently as a product of Aspergillus terreus, is identical to monocolin K.[30] The dihydroderivatives of mevinolin[31] and of compactin[32] have been isolated from Aspergillus terreus and Penecillium citrinum respectively, and resemble in activity their parent compounds.

Enzyme Inhibitors

Since some diseases are associated with excessive or unregulated enzyme activities, enzyme inhibitors from microbial broths have proven valuable pharmacologically. As a result of the work of Umezawa and his group[33] and of others, a large number of extremely potent enzyme inhibitors have been isolated and identified.[34] Some of these microbial inhibitors are described in the following paragraphs.

Inhibitors of 3-hydroxy-3-methylglutaryl-CoA reductase. This rate-limiting enzyme of cholesterol synthesis has been successfully used as an assay to isolate hypocholesterolemic agents. Such agents, e.g., monacolin K (mevinolin), are extremely active in animals[35] and appear to be headed for clinical use.

Inhibitors of dopamine β-hydroxylase, tyrosine hydroxylase and cathechol-O-methyltransferase. Broths screened for activity in these assays have yielded products showing hypotensive activity in animals.[36-38]

Inhibitors of complement. A specific inhibitor of the complement activation cascade is K-76 monocarboxylic acid, a sesquiterpene derivative which is an oxidation product of the natural compound produced by Stachybotrys complementi.[39] It inhibits nephrotoxic nephritis in rats and may be useful in immune complex diseases, allergic diseases and inflammation.

Inhibitors of intestinal glycosidases. Agents inhibiting amylase or invertase might be useful for persons suffering from carbohydrate-dependent diseases such as diabetes, type IV hyperlipoproteinemia and obesity. Such a compound is acarbose (BAYg5421), produced by Actinoplanes sp.[40] Another is product S-AI of Streptomyces diastaticus subsp. amylostaticus, which inhibits α-amylase and glucoamylase but not β-amylase or pullulanase.[41]

A specific inhibitor of Streptococcus mutans dextransucrase has been isolated from a streptomycete. Since dextransucrase is thought to play a role in the initiation of dental caries, the product (ribocitrin) may have application in preventing cariogenicity. Ribocitrin has no antibiotic activity and appears to have no acute toxicity.[42]

Inhibitors of pancreatic esterase. Esterasin, an inhibitor of pancreatic esterase produced by Streptomyces lavendulae, is non-toxic and possesses no antibiotic activity.[43] It suppresses delayed-type hypersensitivity and antibody formation.

Inhibitors of cholinesterase. A compound (I-6123) inhibiting cholinesterase is produced by Aspergillus terreus.[44] This is of interest since known synthetic insecticides are cholinesterase inhibitors.[45]

Protease inhibitors. Due to the possible role of proteases of polymorphonuclear leukocytes in inflammation and carcinogenesis,[46] a large number of potent protease inhibitors have been isolated from streptomycete broths[24]; among the best known are leupeptin,[47] antipain, chymostatin, elastatinal, bestatin and pepstatin. They have been extremely important in studies establishing the role of proteases in various processes, such as carcinogenesis. Furthermore, elastatinal and antipain are active against chemical mutagenesis in bacteria, apparently inhibiting a protease involved in SOS repair.[48] Carboxyprotease inhibitors include pepstatins, pepstanones and hydroxypepstatins. Pepstatin inhibits focus formation by murine sarcoma virus on YH-7 mouse cells and ascitic accumulation in cancer.[49]

Elastase appears to be involved in chronic obstructive lung diseases such as emphysema[50],[51] as well as in pancreatitis, acute arthritis and various inflammations. Elasnin, a nonantibiotic metabolite of Streptomyces noboritoensis, inhibits human granulocyte elastase.[52] Another elastase inhibitor is the peptide, elastatinal, produced by an actinomycete.[53] Serine and thiolprotease inhibitors from microbial broths show anti-inflammatory activity.[24],[51]

Pepstatin, a peptide product of several streptomycetes, is an inhibitor of acid protease (especially pepsin[54]) and has a strong diuretic effect, probably due to inhibition of renin.[49] Other streptomycete-derived pepsin inhibitors include SP-1[55] and the pepsinostreptins; the latter prevent gastric ulceration in rats.[56]

A specific inhibitor of the metallic endopeptidase, thermolysin, is phosphoramidon, a nonantibiotic, nontoxic metabolite of Streptomyces tanashiensis.[57]

Bestatin is a specific inhibitor of aminopeptidase B and leucine aminopeptidase and is produced by Streptomyces olivoretuculi.[58] Since these enzymes are cell surface enzymes in lymphocytes, it was reasoned that they might be involved in the immune response and that compounds binding to such enzymes might be immunomodulators. Bestatin was found to enhance delayed-type hypersensitivity in vivo, activation of peripheral blood lymphocytes by concanavalin A, and the activity of antitumor agents in animals.[24] It increases the number of antibody-forming cells in mice and inhibits slow-growing solid tumors such as the Gardner lymphosarcoma and IMC-carcinoma. In clinical studies, bestatin enhanced immunity in cancer patients[26] and showed a number of other beneficial effects.[34]

Other surface enzymes are alkaline phosphatase and esterase, and microbial inhibitors of these enzymes have immunomodulation activity.[34] Amastatin is an inhibitor of aminopeptidase A and leucine aminopeptidase, and is produced by Streptomyces sp. It is nontoxic and has no antibiotic activity.[59] Forphenicine is an inhibitor of alkaline phosphatase and is produced by Streptomyces fulvoviridis var. acarbodicus.[60] Both products increase the number of antibody-forming cells. Forphenicine enhances delayed-type hypersensitivity and shows activity against solid tumors.[34]

Agents affecting cyclic-AMP (cAMP) levels. Since cAMP levels are altered in cancer, hypertension, asthma, cholera and diabetes, there has been some interest in identifying agents that affect these levels. Of special interest has been cAMP-increasing agents that might increase fat cell lipolysis and bronchodilation.

134

Screening efforts have been directed towards the detection of rabbit brain cAMP phosphodiesterase inhibitors, and several streptomycete products have been isolated.[61,62] A useful agar plate assay using beef brain cAMP phosphodiesterase has been described.[63]

Prolyl-4-hydroxylase inhibitors. Fibrotic collagen accumulation causes fibrotic disease of connective tissue. Lung and liver fibrosis in animal models is prevented by proline analogues. Prolyl-4-hydroxylase is thought to be the target enzyme since there is a hydroxyproline requirement for collagen secretion and thermal stability.[64] Since there are no known nontoxic inhibitors of collagen synthesis, it is of interest that an inhibitor, P-1894B, has been isolated from Streptomyces sp.[65] and found to be identical to antitumor antibiotic vinomycin A_1.[66] P-1894B showed some acute toxicity in rats (LD_{50} = 100-200 mg/kg) when given i.p., but was nontoxic orally.

Ornithine decarboxylase inhibitors. Polyamines such as putrescine, spermine and spermidine play a mysterious but essential role in cell growth, differentiation and multiplication. Interfering with their synthesis could be useful in diseases in which abnormally rapid proliferation of cells occurs.[67] Since L-ornithine decarboxylase is the rate-controlling enzyme of polyamine synthesis in mammalian cells, it is a good target for diseases such as psoriasis, chronic nonsuppurative prostatitis and cancer. An initial screening of microbial broths has led to the isolation from Streptomyces neyagawaensis of two known compounds, dihydrosarkomycin and sarkomycin[68]; the latter is an old antitumor antibiotic.[69]

Angiotensin-converting enzyme. Blood pressure is normally regulated by the renin-angiotensin system. Angiotensinogen (a plasma protein) is cleaved by trypsin and renin to form an inert peptide angiotensin I. This is cleaved by the angiotensin-converting enzyme, a zinc exopeptidase, to angiotensin II, the most powerful vasoconstrictor known. Overproduction of angiotensin II appears to be a major cause of hypertension. A leading synthetic oral hypotensive drug, captopril[70] and newer compounds,[71] act by inhibiting the enzyme. These are small peptide derivatives and it is thus not surprising that an inhibitor has recently been detected in a streptomycete broth.[72]

REFERENCES

1. R. F. Shumard and M. E. Callender, Antimicrob. Agents Chemother. 369 (1968).
2. J. W. Westley, Adv. Appl. Microbiol. 22:177 (1977).

3. M. Chen and M. J. Wolin, _Appl. Environ. Microbiol._ 38:72 (1979).

4. R. W. Burg, B. M. Miller, E. E. Baker, J. Birnbaum, S. A. Currie, R. Hartman, Y.-L. Kong, R. L. Monaghan, G. Olson, I. Putter, J. B. Tunac, H. Wallick, E. O. Stapley, R. Ōiwa, and S. Ōmura, _Antimicrob. Agents Chemother._ 15:361 (1979).

5. L. C. Fritz, C. C. Wang, and A. Gorio, _Proc. Natl. Acad. Sci. USA_ 76:2062 (1979).

6. J. C. Chabala, H. Mrozik, R. L. Tolman, P. Eskola, A. Lusi, L. H. Peterson, M. F. Woods, M. H. Fisher, W. C. Campbell, R. R. Egerton, and D. A. Ostlind, _J. Med. Chem._ 23:1134 (1980).

7. E. O. Stapley and H. B. Woodruff, _in:_ "Trends in Antibiotic Research," p. 154. H. Umezawa, A. L. Demain, T. Hata, and C. R. Hutchinson, eds., Japan Antibiotic Research Association, Tokyo (1982).

8. L. A. Bulla, Jr., D. B. Bechtel, K. J. Kramer, Y. I. Shethna, A. I. Aronson, and P. C. Fitz-James, _CRC Crit. Rev. Microbiol._ 8:147 (1980).

9. P. S. Myers and A. A. Yousten, _Appl. Environ. Microbiol._ 39:1205 (1980).

10. L. A. Bulla, Jr., R. N. Costilow, and E. S. Sharpe, _Adv. Appl. Microbiol._ 23:1 (1978).

11. G. U. Brillinger, _Arch. Microbiol._ 121:71 (1979).

12. T. Leighton, E. Marks, and F. Leighton, _Science_ 213:905 (1981).

13. S. J. Box, M. Cole, and G. H. Yeoman, _Appl. Microbiol._ 26:699 (1973).

14. Y. Takiguchi, H. Mishima, M. Okuda, and M. Terao, _J. Antibiot._ 33:1120 (1980).

15. T. Misato, _in:_ "Pesticide Chemistry in the 20th Century," p. 170. J. R. Plimmer, ed., American Chemical Society, Washington, D.C. (1977).

16. S. Ōmura, Y. Iwai, Y. Takahashi, N. Sadakane, A. Nakagawa, H. Ōiwa, Y. Hasegawa, and T. Ikai, _J. Antibiot._ 32:255 (1979).

17. Y. Takiguchi, H. Yoshikawa, A. Terahara, A. Torikata, and M. Terao, _J. Antibiot._ 32:857 (1979).

18. E. G. Jefferys, _Adv. Appl. Microbiol._ 13:283 (1970).

19. D. Weisinger and J. F. Borel, _Immunobiology_ 156:454 (1979).

20. L. C. Vining and W. A. Taber _in:_ "Economic Microbiology, Vol. 3, Secondary Products of Metabolism," p. 389. A. H. Rose, ed., Academic Press, London (1979).

21. S. Ōmura, Y. Iwai, Y. Suzuki, J. Awaya, Y. Konda, and M. Onda, _J. Antibiot._ 29:797 (1976).

22. P. H. Hidy, R. S. Baldwin, R. L. Greasham, C. L. Keith, and J. R. McMullen, _Adv. Appl. Microbiol._ 22:59 (1977).

23. V. Groupe and R. Donovick, _J. Antibiot._ 30:1080 (1977).

24. T. Aoyagi, M. Ishizuka, T. Takeuchi, and H. Umezawa, _Jap. J. Antibiot._ 30 Suppl, S-121 (1977).

25. J. Itoh, S. Omoto, T. Shomura, N. Nishizawa, S. Miyado, Y.
    Yuda, U. Shibata, and S. Inouye, J. Antibiot. 34:611
    (1981).
26. T. Aoyagi and H. Umezawa, in: "Advances in Biotechnology,
    Vol. 1, Scientific and Engineering Principles," p. 29.
    M. Moo-Young, C. W. Robinson, and C. Vezina, eds.,
    Pergamon Press, Toronto (1981).
27. Y. T. Ninomiya, Y. Yamada, H. Shirai, M. Onitsuka, Y. Suhara,
    and H. B. Maruyama, Chem. Pharm. Bull. 28:3157 (1980).
28. A. Endo, M. Kuroda, and Y. Tsujita, J. Antibiot. 29:1346
    (1976).
29. A. Endo, J. Antibiot. 33:334 (1980).
30. A. W. Alberts, J. Chen, G. Kuron, V. Hunt, J. Huff, C.
    Hoffman, J. Rothrock, M. Lopez, H. Joshua, E. Harris, A.
    Patchett, R. Monaghan, S. Currie, E. Stapley, G. Albers-
    Schönberg, O. Hensens, J. Hirshfield, K. Hoogsten, J.
    Liesch, and J. Springer, Proc. Natl. Acad. Sci. USA
    77:3957 (1980).
31. G. Albers-Schönberg, H. Joshua, M. B. Lopez, O. D. Hensens,
    J. P. Springer, J. Chen, S. Ostrove, C. H. Hoffman, A. W.
    Alberts, and A. A. Patchett, J. Antibiot. 34:507 (1981).
32. Y. K. T. Lam, V. P. Gullo, R. T. Goegelman, D. Jorn, L. Huang,
    C. DeRiso, R. L. Monaghan, and I. Putter, J. Antibiot.
    34:614 (1981).
33. H. Umezawa, "Enzyme Inhibitors of Microbial Origin,"
    University of Tokyo Press, Tokyo (1972).
34. H. Umezawa, in: "Advances in Biotechnology, Vol. 3,
    Fermentation Products," p. XV. C. Vezina and K. Singh,
    eds., Pergamon Press, Toronto (1981).
35. A. Endo in: "Artherosclerosis V," p. 152. A. M. Gotto, Jr.,
    L. C. Smith, and B. Allen, eds., Springer-Verlag, New
    York (1980).
36. H. Iinuma, T. Takeuchi, S. Kondo, M. Matsuzaki, H. Umezawa,
    and M. Ohno, J. Antibiot. 25:497 (1972).
37. H. Chimura, T. Sawa, Y. Kumada, F. Nakamura, M. Matsuzaki, T.
    Takita, T. Takeuchi, and H. Umezawa, J. Antibiot. 26:618
    (1973).
38. H. Umezawa in: "Fermentation Technology Today," p. 401.
    G. Terui, ed., Fermentation Technology Society, Osaka
    (1972).
39. K. Hong, T. Kinoshita, W. Miyazaki, T. Izawa, and K. Inoue,
    J. Immunol. 122:2418 (1979).
40. E. Truscheit, W. Frommer, B. Junge, L. Müller, D. D. Schmidt,
    and W. Wingender, Angew. Chem. Int. Ed. Engl. 20:744
    (1981).
41. S. Murao, K. Ohyama, and S. Ogura, Agr. Biol. Chem. 41:919
    (1977).
42. Y. Okami, M. Takashio, and H. Umezawa, J. Antibiot. 34:344
    (1981).

43. H. Umezawa, T. Aoyagi, T. Hazato, K. Uotani, F. Kojima, M. Hamada, and T. Takeuchi, J. Antibiot. 31:639 (1978).

44. K. Ogata, K. Ueda, T. Nagasawa, and Y. Tani, J. Antibiot. 27:343 (1974).

45. B. H. Chin and N. Spangler, J. Agr. Food Chem. 28:1342 (1980).

46. A. Janoff, Ann. Rev. Med. 23:177 (1972).

47. K. Suzukake, H. Hayashi, M. Hori, and H. Umezawa, J. Antibiot. 33:857 (1980).

48. M. S. Meyn, T. Rossman, and W. Troll, Proc. Natl. Acad. Sci. USA 74:1152 (1977).

49. H. Esumi, S. Sato, and T. Sugimura, J. Antibiot. 31:872 (1978).

50. S. Eriksson, Acta Med. Scand. 203:449 (1978).

51. A. J. Barrett, in: "Enzyme Inductors as Drugs," M. Sandler, ed., Macmillan Press, London (1980).

52. A. Nakagawa, H. Ohno, K. Miyano, and S. Ōmura, J. Org. Chem. 45:3268 (1980).

53. A. Okura, H. Morishima, T. Takita, T. Aoyagi, T. Takeuchi, and H. Umezawa, J. Antibiot. 28:337 (1975).

54. H. Umezawa, T. Miyano, T. Murakami, T. Takita, T. Aoyagi, T. Takeuchi, H. Naganawa, and H. Morishima, J. Antibiot. 26:615 (1973).

55. S. Murao and S. Satoi, Agr. Biol Chem. 34:1265 (1970).

56. T. Kanamaru, T. Asano, H. Torii, J. Kataoka, H. Okazaki, A. Kakinuma, S. Narumi, T. Hirata, K. Gomaibashi, and M. Kanno, J. Takeda Res. Lab. 35:136 (1976).

57. H. Suda, T. Aoyagi, T. Takeuchi, and H. Umezawa, J. Antibiot. 26:621 (1973).

58. H. Umezawa, T. Aoyagi, H. Suda, M. Hamada, and K. Takeuchi, J. Antibiot. 29:97 (1976).

59. T. Aoyagi, H. Tobe, F. Kojima, M. Hamada, K. Takeuchi, and H. Umezawa, J. Antibiot. 31:636 (1978).

60. T. Aoyagi, T. Yamamoto, K. Kojiri, F. Kojima, M. Hamada, T. Takeuchi, and H. Umezawa, J. Antibiot. 31:244 (1978).

61. H. Nakamura, Y. Enomoto, T. Takeuchi, H. Umezawa, and Y. Iitaka, Agr. Biol. Chem. 42:1337 (1978).

62. Y. Furutani, M. Shimada, M. Hamada, T. Takeuchi, and H. Umezawa, Agr. Biol. Chem. 41:989 (1977).

63. P. J. Somers and C. E. Higgens, Appl. Environ. Microbiol. 34:604 (1977).

64. G. C. Fuller, J. Med. Chem. 24:651 (1981).

65. H. Okazaki, K. Ohta, T. Kanamaru, T. Ishimaru, and T. Kishi, J. Antibiot. 34:1355 (1981).

66. S. Ōmura, H. Tanaka, R. Ōiwa, J. Awaya, R. Masuma, and K. Tanaka, J. Antibiot. 30:908 (1977).

67. J. Koch-Weser, P. J. Schechter, P. Bey, C. Danzin, J. R. Fozard, M. J. Jung, P. S. Mamont, N. Seiler, N. J. Prakash, and A. Sjoerdsma, in: "Polyamines in Biology and Medicine," p. 437. D. R. Morris and L. J. Marton, eds., Marcel Dekker, New York (1981).

138

68. A. Fujiwara, Y. Shiomi, K. Suzuki, and M. Fujiwara, <u>Agr.</u>
    <u>Biol. Chem.</u> 42:1435 (1978).
69. H. Umezawa, T. Takeuchi, K. Nitta, Y. Okami, T. Yamamoto, and
    S. Yamaoka, <u>J. Antibiot. Ser. A.</u> 6:101 (1953).
70. M. A. Ondetti and D. W. Cushman, <u>J. Med. Chem.</u> 24:355 (1981).
71. H. Gavras, B. Waeber, I. Gavras, J. Biollaz, H. R. Brunner,
    and R. O. Davies, <u>The Lancet</u> ii:543 (1981).
72. L. Huang, G. Rowin, J. Dunn, R. Sykes, R. Dobna, B. A.
    Mayles, D. M. Gross, J. Liesch, O. D. Hensens, and R. W.
    Burg, Abstract 031, p. 244. <u>Ann. Mtg. Amer. Soc.</u>
    <u>Microbiol.</u> (1983).

BIOCHEMISTRY AND GENETICS OF VITAMIN PRODUCTION

Geneviève C. Barrere

Centre de Recherche de Vitry : Departement Procedes

Rhone Poulenc Sante. France

"Vitamins" is a collective term for certain compounds essential to nutrition. The diet of humans has to contain not only sources of energy, for example carbohydrates, lipids and proteins, and sources of minerals, but also a range of essential compounds called vitamins. These are complex compounds which have no energetic value, and are required only in trace amounts to ensure healthy growth and normal body maintenance in humans. Unlike hormones, which are also required in very small quantities but which are endogeneous metabolites, vitamins must be supplied in the diet.

The term of "Vitamin" was coined by FUNK[1] in 1911. He was the first to isolate an example of such a compound. It was produced in crystalline form, and later become known as thiamine. Since it was an amine, and was essential to the maintenance of life, FUNK called it Vita (life)-amine. It was shown later that not all molecules of the so-called vitamins contain an amine group.

These compounds are essential not only to humans but also to animals, plants and microorganisms. Biosynthetic incompetence which characterizes the nutritional requirement for a vitamin can vary from one species to another. For instance Vitamin C is synthesized by most of the animal species and is a vitamin only for humans, monkeys and guinea-pigs.

For humans the diet has to contain 13 compounds considered as vitamins. Among these 13 compounds, unsaturated fatty acids, such as linoleic acid, are not taken into account. The essential fatty acids are frequently referred to as "Vitamin F". They can be transformed in the cell to arachidonic acid, the deficiency of which causes skin alterations in rats. These fatty acids of the arachidonic group can be transformed into prostaglandins, and have an important role in the regulation of the cellular metabolism.

The daily vitamin requirements of an average man of 70 Kg is shown in Fig.1. This, of course, varies from one adult to another, depending on the climatic conditions, physical activity and age. For example, in northern countries, with little sunshine, people need larger quantities of Vitamin D. Pregnancy is a physiological condition requiring a supplement of all vitamins. Children need to receive a large amount of Vitamins A and D to achieve normal growth. This

VITAMIN	DAILY REQUIREMENT	
Vitamin C	45	mg
Vitamin PP	18	"
Vitamin E	5-15	"
Pantothenic acid	10	"
Vitamin K	4	"
Vitamin B 6	2	"
Vitamin B 2	1.6	"
Vitamin A	1.5	"
Vitamin B 1	1.4	"
Folic acid	0.2	"
Biotin	0.001	"
Vitamin B 12	0.0003	"
Vitamin D	0.0001	"

Fig. 1. Vitamins. Daily Requirements of an average man of 70 kilos. (Food and Nutrition Board-National Research Council, 1974).

maintains a correct Calcium/Phosphate ratio at a stage in development when the fixation of calcium phosphate is very intense.

Nutritionists have calculated the total amount of vitamins consumed by an average man with a 70 year life-span. Total vitamin consumption amounts to a mere 2 Kg, compared with a total food intake over the same period of 12 tons.

Absence or inadequate quantities of one of the vitamins can result in very severe physiological disorders and can even lead to death. We all have heard of Scurvy (the navigator disease), Beri-beri (the Far-east plague) and Pellagra or Rachitis. Each of these diseases have been responsible for innumerable deaths and casualties. Records of these diseases caused through vitamin deficiencies are found in early history. The illnesses were regarded as incurable and could assume the proportions of plagues. It was believed that they were sent by God as a punishment. But as early as medieval times, some people (physicians especially) noticed that the ingestion of certain foods cured these illnesses which could assume the proportions of plagues. These observations led to the following ways of effecting a cure :

      Scurvy : feed fresh fruit and vegetables
   Beri-beri : feed rice envelopes
    Pellagra : feed meat and vegetables
    Rachitis : feed fats

It was not until the beginning of this century that clinical observations and improved knowledge in human nutrition revealed that these disorders, caused by limited diet, were due to the lack of some well-defined substances. These were :

      Scurvy : lack of Vitamin C or Ascorbic acid
   Beri-beri : lack of Vitamin B1 or Thiamine
    Pellagra : lack of Vitamin PP or Nicotinamide
    Rachitis : lack of Vitamin D or Calciferols

To date thirteen of these essential nutritional compounds for humans have been discovered, isolated and studied. An elementary classification is given in Fig. 2.

Structurally, Vitamins represent a very heterogeneous group of products. They are classified according to their solubility properties (i.e. the way in which they partition in the bodies of mammals).

143

WATER-SOLUBLE VITAMINS

Group B	Vitamin B 1	or Thiamine
	Vitamin B 2	or Riboflavin
	Vitamin B 5	or Pantothenic Acid
	Vitamin B 6	or Pyridoxine
	Vitamin B 12	or Cobalamins
	Vitamin B C	or Folic Acid
	Vitamin H	or Biotin
	Vitamin PP	or Nicotinamide
Group C	Vitamin C	or Ascorbic Acid

FAT-SOLUBLE VITAMINS

	Vitamin A	or Retinol
	Vitamin D	or Ergocalciferol and Cholecalciferol
	Vitamin E	or Tocopherol
	Vitamin K	or menaquinones and $\propto$ Phylloquinone

Fig. 2. Elementary classification of human vitamins

Water-Soluble Vitamins

These are the most important in number. With the exception of Vitamin C they all contain nitrogen. Structural analysis reveals their great diversity (Fig. 3).

They are cofactors of important enzymes of the intermediate metabolism.

Their biochemical functions are given in Fig. 4.

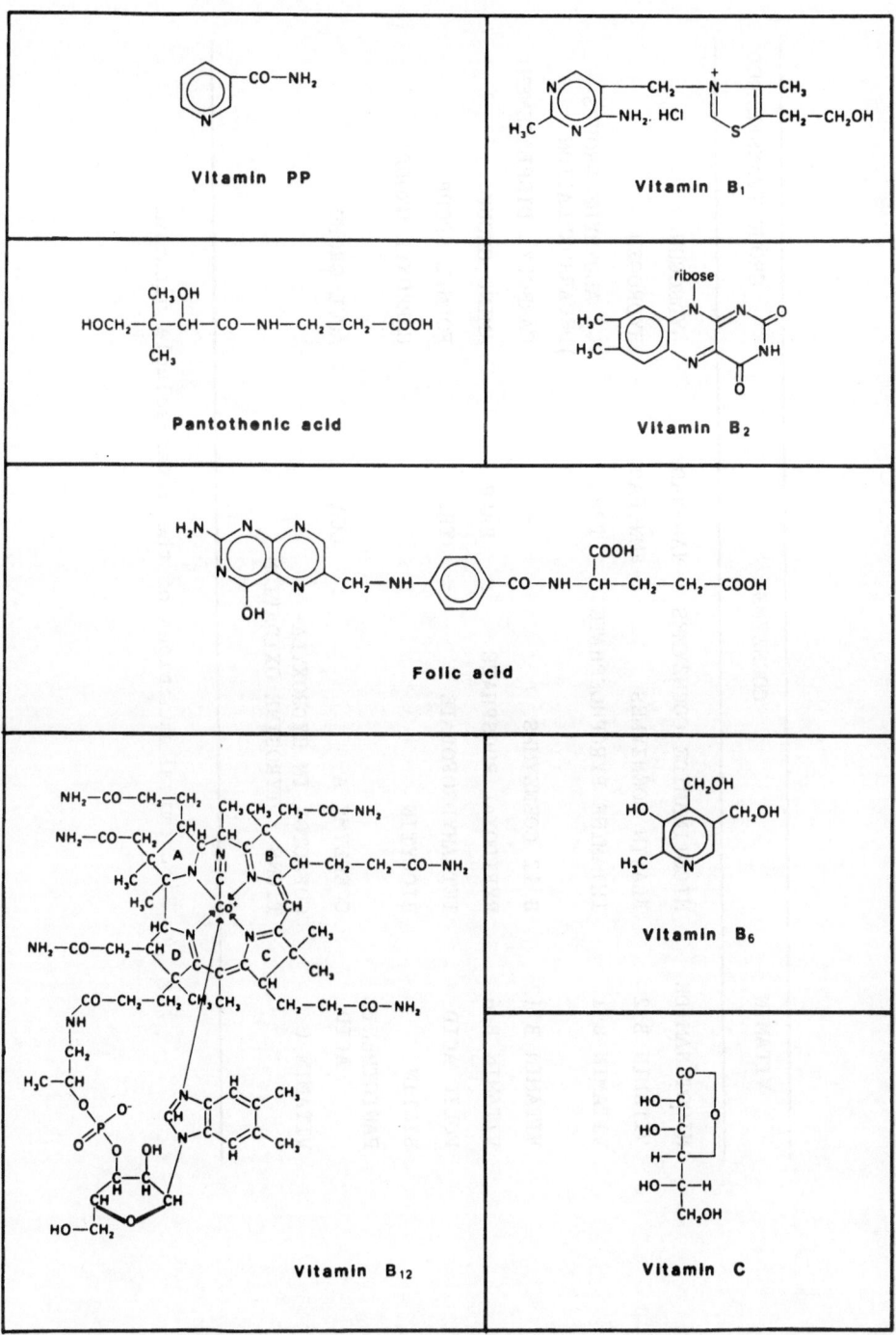

Fig. 3. Chemical structure of the water-soluble vitamins

VITAMIN	COENZYME		GROUP TRANSFERRED
NICOTINAMIDE	NICOTINAMIDE COENZYMES	: NAD-NADP	HYDROGEN
VITAMIN B 2	FLAVIN COENZYMES	: FMN-FAD	HYDROGEN
VITAMIN B 1	THIAMINE PYROPHOSPHATE	: TPP	C_2 ALDEHYDE GROUPS – DECARBOXYLATION
VITAMIN B 12	B 12 COENZYMES	: –	CARBOXYL DISPLACEMENT
VITAMIN B 6	PYRIDOXAL PHOSPHATE	: PALP	AMINO GROUP
FOLIC ACID	TETRAHYDROFOLATE	: THF	FORMYL GROUP
BIOTIN	BIOCYTIN	: –	CARBOXYL GROUP
PANTOTHENIC ACID	COENZYME A	: COA	ACYL GROUP
VITAMIN C	COFACTOR IN HYDROXYLA- TIONS : TYROSINE OXIDATION		

Fig. 4. Biochemical functions of the water-soluble vitamins

Fat-soluble Vitamins

These are considered as lipids and are ultimately derived from isoprenoid building blocks (Fig. 5). The isoprenoid structure is most apparent in Vitamins A,E,K. Vitamin D is a steroid derivative.

They are believed to participate in coenzyme-like functions but for most of them the mode of action is still unknown in molecular terms. Their physiological importance is reported in Fig. 6. These fat-soluble molecules are vitamins only in higher animals.

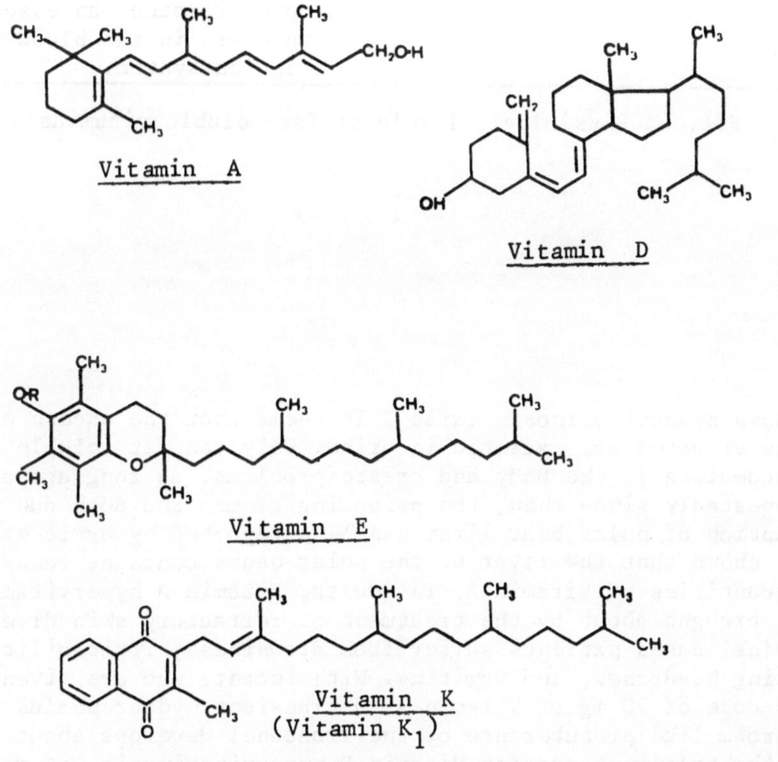

Fig. 5. Chemical structure of the fat soluble vitamins

VITAMIN A	:	VISION	The aldehyde form (RETINAL) combines with a protein-the opsine-to form the rhodopsine that has an important role in vision at night.
VITAMIN D	:	SKELETON	Facilitates the calcium transport in the intestine. Role in the fixation of calcium phosphate on the skeleton.
VITAMIN E	:	SEX.	Cellular antioxydant. Role in reproduction phenomenons
VITAMIN K	:	BLOOD	Role in the synthesis of prothrombine, an essential protein in the blood coagulation system.

Fig. 6. Physiological role of fat-soluble vitamins

Does hypervitaminosis exist ? It seems that the excess of water soluble vitamins are excreted in urine. Only the fat soluble ones can accumulate in the body and create problems. As long ago as 1596, and repeatedly since then, the poisoning of men and dogs due to the consumption of polar bear liver has been reported by artic explorers. It was shown that the liver of the polar bears contains remarkably high quantities of Vitamin A. In adults, Vitamin A hypervitaminosis may be brought about by the treatment of refractory skin diseases. In typical cases patients suffer from drowsiness, irritability, splitting headaches, and vomiting. With infants who are given a single dose of 90 mg of Vitamin A, a transient hydrocephalus with a mushroom like protuberance of the frontanel develops about 12 hours after the treatment. As for Vitamin D hypervitaminosis, it may appear in individuals receiving massive doses of calciferol in the treatment of arthritis or tuberculosis. The basic pathological effect is the precipitation of calcium in various tissues. Renal insufficiency may develop as the result of such calcification in kidneys. Fatal cases result in uremia.

Besides the obvious therapeutic value of Vitamins, the production of these growth-promoting substances has become an important branch of the Food industry.

Purified vitamin preparations can be used as human food additives. During World War II, in the U.S.A., condensed milk was enriched with Vitamin D, bread with Vitamins B1, B2 and PP and margarine with Vitamin A or its precursor β-carotene. In many developed countries where white bread is preferred to wholemeal bread, white bread is supplemented with vitamins and minerals. In 1976, 250 million people consumed vitamin-enriched bread.

Crude vitamin preparations can be used as animal feed additives. The production of eggs, milk and fur by domestic animals can be increased by supplementation of their diet with vitamins. For example 1 g of Vitamin D, 100 g of Vitamin A, 5 g of Vitamin B12, 1 kg of riboflavin are sufficient to fortify 300 to 400 tons of feed, leading to the production of 30-100 tons of swine and poultry.

Vitamin consumption in developed countries is of increasing importance. Vitamin deficiencies exist even in highly developed societies, where the diet is not always as well-balanced as it was thought to be some years ago. An increased requirement is generally recognized in cases of alcoholism or infectious diseases ; Vitamins are prescribed in association with certain therapies, such as the administration of broad-spectrum antibiotics, anticonvulsants and steroid contraceptives.

A rough estimation of the vitamin market is as follows :

- Vitamin C production is over 10,000 tons a year (about 20 % in dollars of the total vitamin market).

- Nicotinamide (Vitamin PP) production approximates to 10,000 tons a year.

- Biotin, Vitamin B12 and Folic acid represent low tonnage. (Vitamin B12 production is about 10 tons a year).

- The other vitamins productions range between 1,000 and 10 000 tons a year.

HOFFMANN-LA ROCHE is omnipresent on this market except for a few vitamins, for example Vitamin B12.

Most vitamins are manufactured synthetically. Although microbial processes for the production of Thiamine, Riboflavin, Biotin, Folic acid, Pantothenic acid, Vitamin B12 and Pyridoxine, have been described in scientific and patent literature, only Vitamin B12 and Riboflavin are produced on a commercial scale by fermentation.

RIBOFLAVIN

Pure cristalline lactoflavin or riboflavin was obtained, from whey, by GYORGY and coworkers[2] in 1933. The structure of this 7,8 dimethyl-10-(D-ribityl) isoalloxazine is shown in Fig. 7.

The world riboflavin production was estimated, in 1977, at about 2,000 tons. The commercial importance of this vitamin has grown considerably in recent years. Its production in 1960-1961 amounted only to 200 tons.

There are no classical specific clinical indications of riboflavin deficiency in man. The first reports of ariboflavinosis date from 1938. Soreness of mouth, lips and nose, genital dermatitis and ocular pathology (hypervascularization of the cornea, conjunctivitis) are the most frequent symptoms observed, but they are not invariate. Therapeutically this vitamin is utilized in the treatment of dermatitis, ocular ailments such as over-sensitivity to light, or hypervascularization of the cornea. It is also used against muscular cramps, migraine, general states of nutritional deficiency, and as a promotor of general good-health.

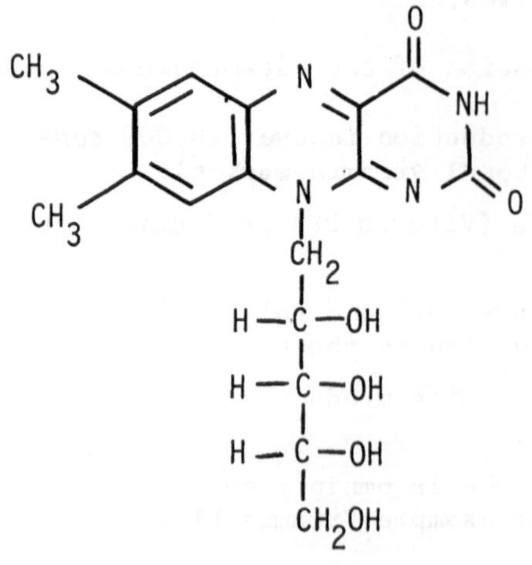

Fig. 7. 7,8-Dimethyl-10-(D-ribityl) isoalloxazine

Commercial production can be achieved by three sorts of processes :

- Complete chemical synthesis
- Complete microbial synthesis by fermentation
- Mixed synthesis : microbial synthesis of ribose followed by chemical transformation.

Today, more than 70 % of the industrial production is made by chemical synthesis from ribose. As none of the known synthesis of D-ribose are simple and economical, the pentose sugar is mainly obtained by fermentation processes. TAKEDA Chemical Ind. [3,4,5,6] utilizing very high ribose producing mutants of *B. subtilis* and other *Bacillus* have reported production yields as high as 70 g/1.

Most of the remaining 30 % of the riboflavin production is of microbial origin. The only current fermentation process was introduced at MERCK and Co.

Riboflavin is synthesized by most higher plants and various microorganisms. Higher animals are unable to synthesize it ; their vitamin requirement must be supplied by the microflora of their gastro-intestinal tract or by their food.

In some of the producing microorganisms, the amount synthesized far exceed the metabolic requirement. In these organisms, some of the vitamin is excreted into the medium.

According to DEMAIN[7], these natural overproducers can be divided in three groups : weak overproducers (i) - moderate overproducers (ii) - strong overproducers (iii).

(i) Weak overproducers : *Clostridia* species are representative of weak overproducers. *Clostridium acetobutylicum*, used in the industrial production of acetone and butanol was, in 1933, the first organism reported to produce large quantities of riboflavin. The residues left after the distillation of the solvents were used in the years immediately prior to World War II as a commercial source of this vitamin[8]. It was soon discovered that the synthesis of riboflavin by these anaerobes was very sensitive to various ions and especially to iron. Sensitivity of *Clostridia* to iron is indicated in Fig. 8. Yields of nearly 100 mg/1 (in 4 days) were obtained by controlling the inorganic iron content of media containing whey or cereal grain mashes[9].

Microorganism	Optimum iron concentration (mg/l)
Clostridia acetobutylicum	1 – 3
Candida flarei	0.04 – 0.06
Eremethecium ashbyii	not critical
Ashbya gossypii	not critical

Fig. 8. Optimum iron concentration for riboflavin overproduction

(ii) Moderate overproducers : These include yeasts, especially *Candida* species. *C. guilliermondii* and *C. flarei* are the most flavinogenic species. They have been used only on a semi-commercial scale ; their high iron sensitivity (Fig 8), a hundred times higher than that of *Clostridia* species, do not favour their commercial utilization. The complex fermentation media used must be freed of excess iron and glass or plastic coated fermentors are necessary. In laboratory conditions, LEVINE and Coworkers [10] obtained with *C. flarei* in 7 days a riboflavin yield of 600 mg/l. They used an unsterilized medium ; in a sterilized medium, addition of glycine and asparagine is necessary. The initial pH of the unsterilized medium was adjusted below 5.0. The medium contained urea as sole nitrogen source and an optimal concentration of iron.

(iii) Strong overproducers : although fungi in general do not over-produce riboflavin, two yeast-like fungi, the ascomycetes *Eremethecium ashbyii* and *Ashbya gossypii* are the most active producers known today. Both are plant pathogens which commonly infect the cotton plant. They differ essentially in their reproduction-cycle characteristics ; the former is an heterothallic species ; the latter is homothallic. They both form needle-shaped ascospores.

Neither of the organisms appear affected by the iron content of the fermentation medium (Fig 8) : this is a distinct advantage for industrial processes. Fermentation by either organism is aerobic, and optimal at 26-28 °C. Their fermentation characteristics are quite similar.

The riboflavin synthesizing capabilities of *E. ashbyii* were first noted in 1935 by GUILLIERMOND and coworkers[11] and commercial processes were used in the early 1940's. The highest yields were reported by MOSS and KLEIN[12] : 2480 mg/l in a medium containing lentils and molasses (molasses were fed to control pH at about 7.0). *Eremethecium ashbyii* was rather subject to genetic variation. Maintenance of high flavinogenic strains was a major problem, as low producing substrains readily appeared in the population of high producing mutants. It is not until 1957 that an efficient method of maintenance was described[13].

In 1946 WICKERMAN and coworkers[14] shown that under appropriate conditions selected strains of *Ashbya gossypii* were capable of producing substantial amounts of riboflavin. The first yields reported of 300-500 mg/l were below those given for *E. ashbyii*, but because *A. gossypii* appeared genetically much more stable, processes employing the latter gained ground rapidly. The sole fermentation process in commercial use today utilizes *A. gossypii*.

The composition of the culture medium and the temperature of incubation are the most important factors influencing riboflavin production by *A. gossypii*. Cultural conditions such as inoculum development and sterilization were found to have also an important bearing upon the riboflavin production by this organism.

TANNER and coworkers[15] described a production medium containing carbohydrates (glucose, sucrose, maltose) as carbon source, and corn-steep liquors, distiller's solubles as well as animal stick liquor, tankage or meat scraps as nitrogen source. Greater yields were obtained when the animal wastes were subjected to papain or trypsin digestion before their incorporation into the medium. Proteins of animal origin were more effective than plant products. With this medium, the use of small quantities (0.5 - 1 % v/v) of young inocula (18-24 h old) and a minimum sterilization time allow yields of 700 mg/l in 8 days of incubation. Pilot plant fermentations yielded 500 to 850 mg/l. Further improvements have resulted from the utilization reported by MALZAHN and coworkers[16] of oil (corn oil, soybean oil, linseed oil. etc) as the carbon source, of an enzymatically degradated collageneous protein as the nitrogen source, together with additional substances such as corn-steep liquors, distillers' solubles or brewers'yeast. Collagen is the main constituent of animal connective tissue. Proteins characterized as collagens are exemplified by gelatin, hide or bone glue, collagen concentrates (liquors from the lard rendering processes). Although collagen has a high glycine content, addition of glycine (0.05-0.4 %) to the medium was beneficial. In this process the prolonged sterilization of the medium improved the vitamin yield. In a 7 days cycle, production yields were reported to reach 4 g/l. With a similar process and additions of 2.5 % of hide or bone glue twice during the 5 days cycle CZCZESNIAK[17] obtained, in laboratory conditions, yields as high as 6.4 g/l.

The optimum production temperature of 28 °C is below the optimum temperature for growth. As shown by KAPLAN and DEMAIN[18] high temperature (37 °C) imposed during growth is especially inhibitory to subsequent riboflavin formation (vitamin production begins after growth completion). If growth is allowed to take place at 28 °C, subsequent riboflavin formation is the same regardless of the subsequent fermentation temperature, within the range 28-37°C. This suggests that growth at low temperature derepresses the formation of the system required for riboflavin overproduction.

Successful strain improvement was reported in 1963. MALZAHN and coworkers[19] utilized the classic mutation-selection technique : after UV irradiation or treatment with uranylnitrate, nitrogen mustard or ethyleneamine, they selected colonies of increased pigmentation when grown on a malt-yeast extract agar medium. One of these mutants produced 3.6 g/l compared to 1.4 g/l for the parental strain. Later on, improved mutants were obtained at MERCK and Co. as spontaneous mutants or after exposure to N-nitroso N-methyl urethane, ICR 191 and ethylmethanesulfonate. According to LAGO and KAPLAN[20] yields have been increased over 11-fold in optimized fermentation conditions.

In complex media, overproduction really begins after growth has ceased. During growth the pH falls from neutrality to pH 4.0-5.0. This fall in pH is due to the formation of acids such as pyruvate or acetate. As growth slows down, these acids are depleted and the pH rises. Only then does riboflavin production start. Crystals of the vitamin appear inside the hyphae. They can even be observed outside the mycelium. An autolytic phase follows riboflavin production. The characteristic drop in pH which occurs in complex media is not invariate. In chemically defined media, pH remains constant and overproduction starts before growth completion.

In carbohydrate-based processes riboflavin production does not begin until the residual sugar concentration falls below 20 g/l. If the sugar is added after growth it does not repress riboflavin formation. In oil-based fermentations oil is still present in large quantities at the start of vitamin production.

There is a clear relationship between sporulation and riboflavin production. Sporulation - when it does occur - takes place during the riboflavin overproduction phase. Conditions that lead to a poor overproduction of riboflavin were found to inhibit sporulation ; nonproducing mutants are asporogeneous and overproducers sporulate well.

In oil media devoid of sugar, riboflavin formation is high and there is a complete lack of sporulation. Cells have a swollen appearance, are extremely vacuolated and they are filled with yellow riboflavin crystals. In sugar fermentations, where the vitamin yields are lower, a considerable fraction of the cells sporulate. The presence of riboflavin cannot be detected in sporulating cells. All these observations suggest that riboflavin over-synthesis is a branch pathway from the one leading to sporulation.

Studies on riboflavin requiring mutants of *S. cerevisiae* by BACHER and LINGENS[21], isolation of various intermediates and enzymes, together with recent studies at the M.I.T. (Massachusetts Institute of Technology)[22], have led HOLLANDER and coworkers[23] to propose the biosynthetic pathway given in Fig. 9.

It has long been known that riboflavin originates from a guanine derivative. Purines stimulate the vitamin production in all the organisms studied, and glycine, a known guanine precursor, stimulates its overproduction in *Candida* species and *A. gossypii*. The purine ring of the guanine compound, with the exception of carbon 8, is incorporated intact into the isoalloxazine ring of riboflavin. The guanine derivative is currently believed to be guanosine 5'-triphosphate (GTP).

This precursor (structure ①) is converted into 2,5 diamino-6-oxy-4(5'phosphoribosylamino)pyrimidine or PRP (structure ②) by the GTP-cyclohydrolase II[24]. C8 is lost as a one carbon unit and found as free formic acid.

PRP undergoes deamination at C2 of the ring and reduction of the ribosylamino group to lead to 5-amino-2,6-dioxy-4 (5'phosphoribitylamino) pyrimidine or ADRAP-5'P (structure ④). It is also called 4-ribitylamino-5aminouracil or R DAU. In yeast and *A. gossypii* genetic and enzymatic evidences indicate that reduction precedes deamination[21,22]. In *E. coli*, enzymatic studies revealed that deamination occurs first[25]. It has been shown that in *B. subtilis*, formation of the ribityl chain passes through a stage of ribulose followed by its reduction to ribityl (ribosyl-→ribulosyl-→ribityl-).

The next intermediate in the pathway [6,7-dimethyl-8-ribityl-lumazine (DMRL or DRL) - (structure ⑤)] results from the addition of a "C4 unit" to the ADRAP-5'P precursor. The nature and origin of this "C4 unit" has recently been elucidated [23] : the source of the "C4 unit" appears to be the 5 carbon ribityl group of ADRAP-5'P ; after dephosphorylation, two molecules of ADRAP-5'P are involved, one as a donor of the "C4 unit", the other as the "C4 unit" acceptor.

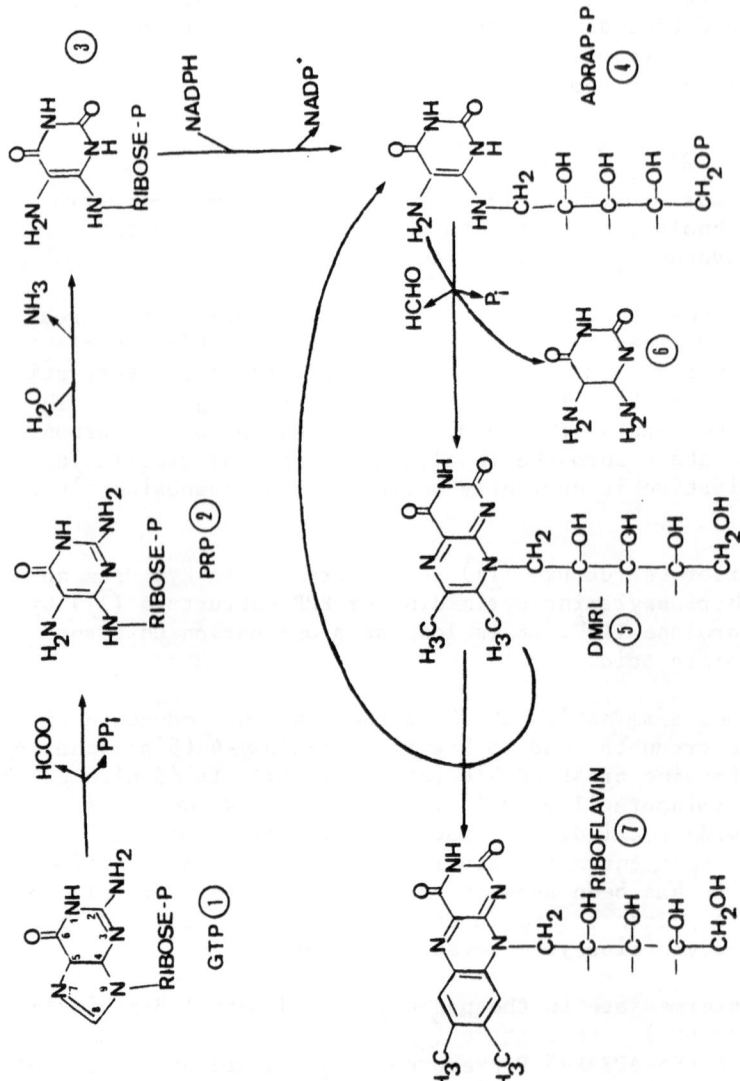

Fig. 9. Riboflavin biosynthetic pathway

GTP:Guanosine 5'-triphosphate   PRP: 2,5-diamino-6-oxy 4(5' phosphoribosylamino)pyrimidine   ADRAP-P or R-DAU: 5-amino-2,6-dioxy 4 (5' phosphori-
                                bitylamino)pyrimidine   DMRL: 6,7-dimethyl-8-ribityllumazine

This transformation may involve the formation of 6-methyl-7-dihydro-xyethyl-8-ribityllumazine or MERL as an intermediate, together with that of diaminouracil (structure ⑥) and formaldehyde :

2 (ADRAP-5'P) ⟶ Pi + FORMALDEHYDE + DIAMINOURACIL + DMRL

Then two molecules of DMRL react to give one molecule of riboflavin (structure ⑦) : One molecule of DMRL gives its recently acquired "C4 unit" to the other molecule of DMRL, thus producing a molecule of riboflavin, and one of ADRAP :

2 DMRL ⟶ RIBOFLAVIN + ADRAP

This reaction is catalyzed by the riboflavin-synthetase which has two sites for DMRL, a donor site and an acceptor site[26].

Riboflavin-deficient mutants of *S. cerevisiae* were classified in 5 biochemical groups according to their accumulation products[27]. Genetically the mutants fall in 6 complementation groups ; DMRL formation is carried out by the products of two genes[28]. No linkage was found between the genes. In *E. coli*, studies recently published found three linkage groups[29]. In *B. subtilis*, the group of genes responsible for riboflavin biosynthesis is a component of one linkage group located on the terminal part of the chromosome[30].

It appears that, in iron sensitive species, riboflavin synthesis is repressed by an iron flavoprotein. In that case, growth in iron deficient medium would produce little or no repressor, thus derepressing the vitamin biosynthesis. In *Candida guilliermondii*, it was shown that oversynthesis of the vitamin in low-iron content media is associated with a derepression of the enzymes of the pathway. This repression effect exerted by an iron containing protein has clearly been shown for the first and last enzymes of the pathway, that is GTP-cyclohydrolase and riboflavin synthetase[31]. According to DEMAIN[7] two possibilities can explain the iron insensitivity of riboflavin production in the two ascomycetes *E.ashbyii* and *A. gossypii*. One possibility is that these species are constitutive with respect to riboflavin synthesizing enzymes. The other one is that empirical conditions that stimulate the vitamin oversynthesis also inhibit the formation of the repressor. Temperature could be one such condition ; growth at 28 °C may not allow the synthesis of the repressor normally synthesized at 37 °C. As cells grown at 28 °C continue to produce riboflavin when shifted up to 37 °C the flavinogenic enzymes must be synthesized early and remain stable through the overproduction phase. This is confirmed by the observation that subinhibitory (for growth) levels of azaguanine and chloramphenicol inhibit riboflavin synthesis only when added during growth[32].

Riboflavin oversynthesis appears to be regulated by the vitamin itself or the flavin coenzymes. In *Candida guilliermondii*, GTP-cyclohydrolase not only has its synthesis regulated by iron but its activity is under the control of feed-back inhibition by the flavin coenzyme FAD[33]. It was shown that it is the adenyl part of the molecule which is effective as an inhibitor. The linkage group in *Bacillus subtilis* forms an operon utilizing FAD, FMN, and riboflavin as effectors and possesses at least two operators [34,35,36,37]. Regulation is accomplished by the regulator protein coded by gene *rib C* unlinked to the riboflavin structural genes. At least two operator-promotor regions participate in this regulation :

- the operator region *rib O* which controls the expression of the genes of the late stages of the biosynthetic pathway from 4-ribulosoamino 5-aminouracil [or 5-amino-2,6-dioxy-4 (5'-phosphoribulosylamino)pyrimidine] to riboflavin. The formation of the enzymes involved in these late stages are repressed by low levels of riboflavin (0.5 μg/ml), or by FAD or FMN.

- the operator region *rib O_e* which controls the expression of the early genes of the pathway, from GTP to 4-ribulosoamino 5-aminouracil. It is assumed that for interacting with *rib O_e* the regulator protein should form a complex with a special effector. This effector may be the 4-ribitylamino-5-aminouracil or ADRAP itself or one of its precursor in the pathway. However, at high concentrations (20 μg/ml), riboflavin itself can also serve as an effector for this operator-promotor ; the regulator protein-riboflavin complex evidently has a reduced affinity for the *rib O_e* region. In cells with a wild type regulation, the synthesis of all the enzymes is regulated by riboflavin in coordination. When the *rib O* operator is shut off under the action of the regulator protein-riboflavin complex, the effector of *rib O_e* (an intermediate of the pathway prior to the late stages of biosynthesis) accumulates and performs its regulatory role on the synthesis of the first enzymes of the pathway. 4-ribitylamino-5-aminouracil (ADRAP) is formed from GTP but also from DMRL ; probably this compound is used again in the biosynthetic pathway. This two steps regulation mechanism brings this intermediate (ADRAP) content to equilibrium with the content of the remaining intermediate products and thereby prevents overconsumption of GTP.

VITAMIN B12

The discovery of this vitamin resulted from attempts to cure pernicious anemia. In 1926, two american physicians, MINOT and MURPHY[38] showed that this anemia could be therapeutically controlled by raw liver. Twenty-two years later the "liver factor" was isolated simultaneously by several independant teams. FOLKERS and coworkers[39] at MERCK and Co. succeeded in isolating the first red crystals from liver and also from a streptomycin fermentation broth.

Eight months later SMITH and PARKER[40] from GLAXO Laboratories reported the isolation of 1 g of the crystalline compound from 4 tons of ox liver by partition chromatography. It is not until 1955 that Dorothy HODGKIN[41] elucidated its structure. Totally synthetic crystals were not obtained before the 1970's.

In 1980, the world production was estimated at 12000 kilos, 54 % for human utilization (pharmaceutical and dietary supplements), the remaining 46 % used as a growth factor in animal feed. Most of this production is manufactured by three companies ; MERCK and Co. GLAXO Laboratories and RHONE-POULENC. The latter supplies more than 50 % of the world market.

Therapeutically vitamin B12 is used in microgram doses for the treatment of pernicious anemia, general states of nutritional deficiency, peripheral neuritis and nevralgia. Because of its lack of toxicity it is widely used in all sorts of chronic incurable afflictions such as arthritis and psoriasis. In many countries it is also used as a remedy for tiredness and miscellaneous aches and pain.

In industrial husbandry where the diet is composed only of vegetable material, vitamin B12 is incorporated in the animal feed at a dose of 10-15 milligrams per ton.

It is now well established that only certain microorganisms have the ability to synthesize the entire cobalamin molecule. A wide range of bacteria, including *Streptomyces*, can produce vitamin B12, but no fungi or yeasts have been reported to do so. The vitamin is not present to a large extent in higher plants and the small amounts found are assumed to be due to bacterial synthesis. In animals, Vitamin B12 is provided for either from their digestive tract microflora or from their food. Humans are completely dependent on food consumption for vitamin B12 since we appear unable to utilize cobalamins synthesized by the bacteria of the gut.

Molecular Structure

Vitamin B12 was the first organometallic compound isolated from biological systems. It is an extraordinarily complex non-polymeric molecule. It is, in fact, the most complex non-polymeric organic compound in nature. The structure of this molecule is shown in Fig.10.

The molecule is schematically composed of two planar (or nearly planar) cyclic regions and a linear region. The metal compound - a trivalent cobalt atom - is linked to one of the cyclic region, a macrocyclic ring system closely related to the porphyrin moiety of heme (the non-protein component of cytochromes). This ring system (the corrin nucleus) like that of porphyrin, is a tetrapyrrole structure, but instead of having its four rings linked by methene groups, its

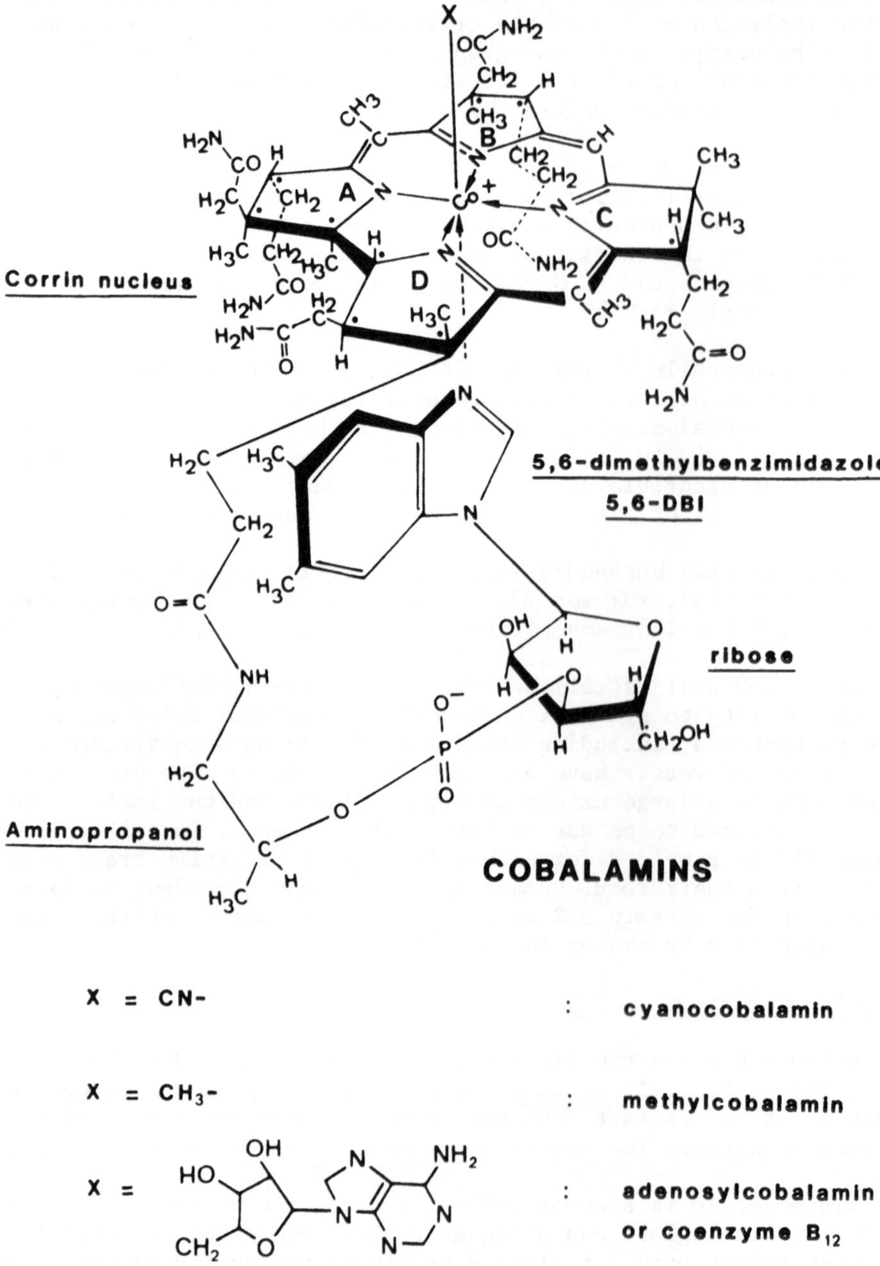

COBALAMINS

X	=	CN-	: cyanocobalamin
X	=	CH$_3$-	: methylcobalamin
X	=		: adenosylcobalamin or coenzyme B$_{12}$

*modified from "Vitamin B$_{12}$". B. Zagalak, W. Friedrich, Edts. Walter de Gruyter and Co. Berlin. N.Y. (1979)

Fig. 10. Structure of cobalamins*

rings A and D are directly linked. The second ring system is a
nitrogen base -the 5,6-dimethylbenzimidazole or DBI. It is linked
to the first ring system by an heterogeneous side chain. The resulting "nucleotide loop" consists in a D-1-amino-2-propanol group
(isopropylamine or IPA) esterified to the phosphate of a 3-mononu-
cleotide linked to its base, the 5-6 DBI, by a N-α-glycosidic bound.
This structure is not only complex but it also possesses some unusual
features : the corrin nucleus was unknown in organic chemistry
before the vitamin B12 structure came to light. Few naturally-exis-
ting ribose-3-phosphate compounds have been discovered α-glycosidic
bounds are very uncommon.

The cobalt atom has six coordinate bounds : four to the pyrrole
nitrogens, one to the N3 of the 5-6 DBI, and the sixth to an "upper
ligand", the nature of which can vary. In commercial vitamin B12
(or cyanocobalamin) the ligand is a -CN group that is an artifact
of the isolation process. *In vivo* the more common ligands are a
deoxyadenosyl group (coenzyme B12), a methyl group (methylcobalamin)
or an hydroxo group (hydroxocobalamin). Apart from these molecules,
collectively known as cobalamins, there are many more instances of
corrin molecules or corrinoids since the side chain can differ in
its nucleotide base. This can be completely lost (if the whole nucleotide is absent, we have the Factor B), or the 5-6 DBI can be
replaced by, for example, adenine (pseudo-vitamin B12), guanine
(Factor C), 2-methyladenine (Factor A) etc...which can be active in
some microorganisms but are inactive in humans. So, when we talk of
vitamin B12 we generally refer to a family of structurally-related
compounds. Among all these "vitamin B12 like" substances, only deoxy-
adenosylcobalamin and methylcobalamin are really active at the cellular level and are involved as cofactors in ten or a dozen enzymatic reactions.

Biosynthesis

Vitamin B12 biosynthesis proceeds from the successive formation
of :
- the porphyrin nucleus
- the corrin nucleus
- the cobalamins.

The biosynthetic pathway is shown in Fig. 11.
Today the porphyrin pathway is well defined. The four successive
reactions leading from the linear molecule- δ-aminolevulinic acid-to
the macrocyclic intermediate-uroporphyrin III-have been studied in
various microorganisms. Enzymology of this part of the vitamin B12
pathway is well-known, especially in what concerns the first two
steps. In contrast, the successive steps leading from uroporphyrin
III to cobyrinic acid, including the corrin nucleus formation, are

still partly unknown. The detection, in the past five years, of several new pigmented intermediates- the methylcorriphyrins- threw some light on this puzzling part of the vitamin pathway. Their isolation allowed the identification of the sequence of the successive methylations of the corrin nucleus. It also proved that the insertion of the cobalt and the formation of the corrin structure by loss of the C 20 carbon take place after the methylations have occurred. It is well established that S-adenosylmethionine is the methylating agent of all these methylations. The enzymology of these reactions, nevertheless, is yet to be understood. For instance, one still wonders if these methylations are achieved by independant enzymes or by a multienzymatic system. The demonstration, in 1977, that dimethylcorriphyrin was absolutely identical to sirohydrochlorin (the siroheme moiety of the microbial sulfite reductase) revealed that this intermediate and perhaps other corriphyrins have a vitamin-independent biological effect. We know that organisms that do not synthesize bitamin B12 do form urophorphyrin III. It was recently shown that some of them also form dimethylcorriphyrin.

The formation of cobalamin involves :

- amidation of the 7 carboxylic acid groups of the corrin nucleus
- incorporation of the isopropylamine residue
- activation of the cobinamide by GTP to form cobinamide-GDP
- insertion of the base nucleotide

The isopropylamine part of the "nucleotide loop" has been shown to derive from threonine by decarboxylation of the amino acid itself or a relative derivative. 5-6 DBI is incorporated into the molecule as the 5'nucleotide. The nucleotide-ribose is esterified at C3 with the phosphate of GDP-cobinamide to form the vitamin 5'phosphate. This phosphate is then removed by a specific phosphatase.
The introduction of the "upper cobalt ligand" seems to occur soon after the formation of cobyrinic acid. In *Propionibacterium shermanii* all known corrinoids more amidated than (but excluding) cobyrinic acid, are in the 5' adenosyl form.
Aerobic and aerotolerant bacteria synthesize their 5-6 DBI moiety from riboflavin via FMN (flavin mononucleotide) ; C'1 of riboflavin is transformed in C2 of 5-6 DBI. This synthesis requires oxygen. It has recently been reported[42,43] than anaerobic microorganisms form the 5-6 DBI moiety by a completely different pathway using glycine and methionine : the two methyl groups of 5-6 DBI are formed from methionine (via S-adenosylmethionine), and glycine is incorporated as a building block. It has also been shown that δ-aminolevulinic acid resulting from the condensation of glycine and succinyl-COA in aerobic species, is formed in anaerobes from glutamate as it has recently been found in plants[42,44].

Regulation studies showed that vitamin B12 influenced none
of the porphyrin pathway enzymes. In contrast, heme affects the
enzymatic system common to both corrinoids and porphyrins biosynthesis. According to BYKHOVSKI[45] (Fig. 11), vitamin B12 synthesis
in *Propionibacterium shermanii* is regulated at the monomethylcorri-
phyrin formation step. This regulation is exerted by cobalamins
such as cyanocobalamin or adenosylcobalamin and also by analogs in
which 5-6 DBI is replaced by, for instance, adenine or methyladenine.
Only the nucleotide free analog (factor B) has no regulatory effect.
We describe later the effects of such regulation on the fermentation of *Propionibacteria* species.

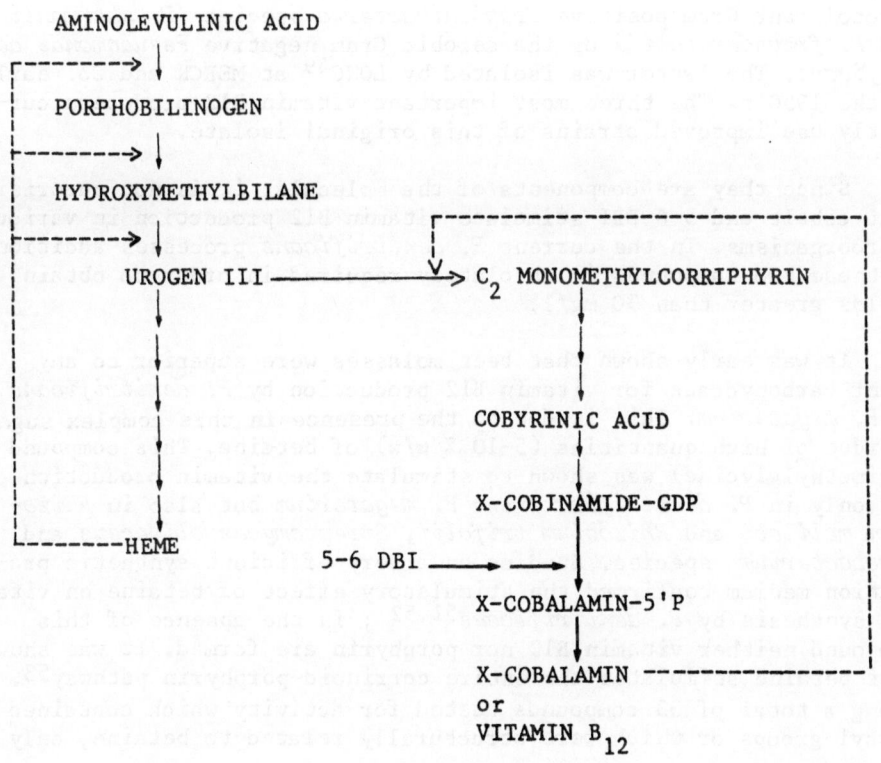

Fig. 11. Regulation of the vitamin B 12 Biosynthetic pathway

The industrial production of vitamin B12 started immediately after the first reports of its isolation. Recovery from activated sewage sludge containing a useful amount of vitamin had been investigated only on a semi-commercial scale. The industrial vitamin was first recovered as by-products of the then current bacterial fermentations : streptomycin and chlortetracycline. Enhanced production was obtained by addition of cobalt salts to the fermentation media. Soon after, the search for more specific and efficient microorganisms led to the description of at least a hundred microbial processes. Only half a dozen of them have been used on a commercial scale. Organisms used include *Streptomyces olivaceus*[46], *Bacillus megaterium*[47], an unidentified *Streptomycete*[48], *Pseudomonas denitrificans*[49] and *Propionibacteria*[50]. In the 1970's, vitamin B12 production by various alcohol or hydrocarbon-utilizing microorganisms was undergoing laboratory-scale study. The occurence of the international oil crisis slowed down these studies.

Today the only commercial production is still by fermentation. Industrial production of such a complex structure by chemical means is practically impossible. The current processes involve either the aerotolerant Gram positive *Propionibacteria* species *(P. shermanii* and *P. freundereichii)* or the aerobic Gram negative *Pseudomonas denitrificans*. The latter was isolated by LONG[49] at MERCK and Co. early in the 1950's. The three most important vitamin B12 producers currently use improved strains of this original isolate.

Since they are components of the molecule, it is not surprising that cobalt and 5-6 DBI stimulate vitamin B12 production in various microorganisms. In the current *P. denitrificans* processes addition of these two compounds is absolutely required in order to obtain yields greater than 50 mg/l.

It was early shown that beet molasses were superior to any other carbohydrate for vitamin B12 production by *P. denitrificans* or *B. megaterium*. This is due to the presence in this complex sugar residue of high quantities (5-10 % w/w) of betaine. This compound (trimethylglycine) was shown to stimulate the vitamin production not only in *P. denitrificans* and *B. megaterium* but also in *Rhizobium meliloti* and *Rhizobium trifolii*, *Streptomyces olivaceus* and *Agrobacterium* species. Studies on a very efficient synthetic production medium confirmed the stimulatory effect of betaine on vitamin B12 synthesis by *P. denitrificans*[51,52] ; in the absence of this compound neither vitamin B12 nor porphyrin are formed. It was shown that betaine stimulates the entire corrinoid-porphyrin pathway[53]. Among a total of 33 compounds tested for activity which contained methyl groups or which were structurally related to betaine, only

betaine and to a less extent choline stimulated Vitamin B12 production. The exact role of betaine is not well understood. Although
it is a methyl donor, betaine is not at the origin of the six methyl
groups of the corrin nucleus. It does not require to undergo metabolism through its known degradative pathway in order to be stimulatory. Betaine has a noticeable effect on the microstructure of
Pseudomonas denitrificans[54] : Electron-micrographs of 72 h cultures
showed that cells grown in the presence of betaine have fine undu-
lations of the cell envelope with several electron dense lamina,
compared with the smooth cell envelope with only two distinct lamina
when grown without betaine. It is possible that betaine increases
permeability of the cells during the period of vitamin B12 overproduction.

In *Propionibacterium shermanii*, ammonium ion has been reported
to have a stimulatory effect[55]. In the absence of ammonium salt,
P. shermanii cells do not produce vitamin B12 even when δ-aminolevuli-
nate is added. Only glutamine and asparagine can replace ammonium
salt.
In order to obtain a high level of production in *P. shermanii* fermentations, part of the process must be run in anaerobic conditions.
The SPEEDIE and HULL[50] process has been among the most productive
for *Propionibacteria* described in the scientific literature. In this
batch process, fermentation in a glucose (or lactose)- corn-steep
liquor medium is carried out under anaerobic conditions until the
completion of growth. Anaerobiosis is obtained by passing a non-
oxidizing gas such as nitrogen or carbon dioxide through the medium.
After 70-80 hours, oxygen or an oxygen-containing gas is introduced
into the medium. Carbohydrate is added if its residual concentration
is too low, and the fermentation is allowed to proceed for up to
120 hours. When oxygen is introduced, it is best to maintain micro-
aerobic, rather than fully aerobic, conditions since excessive oxy-
genation tends to reduce the yield of vitamin. In the continuous
process of SPEEDIE and HULL[50] the fermentation is carried out in
two zones. In the first one, anaerobic conditions are maintained.
Cells containing medium is continuously transferred to a second
aerated zone before being withdrawn for vitamin extraction. Addition
of 5-6 DBI can take place during the aerobic phase. If added earlier,
during the anaerobic development, it results in a lower production
of the vitamin. Why are these two-steps processes so productive ?
Why does the early addition of 5-6 DBI reduce the vitamin production ? We mentioned earlier that the formation of 5-6 DBI requires
oxygen so that in anaerobiosis only incomplete cobalamins can be
synthesized. SPEEDIE and HULL reported that during these initial
conditions, the formation of factor B (cobalamin without the 5-6 DBI
nucleotide) is stimulated. Because these incomplete cobalamins have
no regulatory effect on their own synthesis they accumulate.

B. YONGSMITH ; K. SONOMOTO ; A. TANAKA ; S. FUKUI

Eur. J. Appl. Microbiol. Biotechnol (1982) 16 : 70-74

{ · 5 g   Wet cells in late log phase
{ · 1 g   Urethane propolymer PU-9
    · Resulting resin gel cut in 5 mm cube
    · 50 ml optimized medium (5,6-DBI,Co++)
    · Static cultivation at 30 °C
    · 3 days

Fig. 12. Immobilized *Propionibacterium* sp. cells.

During the following aerated stage they are transformed into complete cobalamins. The early addition of the precursor, 5-6 DBI, results in the early formation of cobalamins that specifically repress the vitamin biosynthesis, and leads to a reduced production.

Recently YONGSMITH and coworkers[56] have reported the ability of immobilized cells of *Propionibacteria* to form vitamin B 12. When cells are immobilized as it is shown in Fig. 12, 10 g wet cells can produce 12.4 μg/ml within 6 days. When the medium is changed every three days, the entrapped cells can carry 5-6 successive batches of medium (total duration : 15-18 days) keeping their capability of production (50 % drop in production after 5-6 batches) ; 180 μg of vitamin per gram of wet cells are produced in 18 days. The excreted vitamin is mainly found in the hydroxo form.

Strain improvement has mainly been achieved in the research laboratories of commercial producers and only very few publications are available. Russian teams described the positive effect of various mutagenic agents on the isolation of vitamin overproducers of *Propionibacterium shermanii*[57-58]. In the past year, strain improvement has proceeded mainly through mutation and selection. By this means, over a 30 year period, *P. denitrificans* strains were obtained that showed a greater than 50-fold increase in productivity relative to the performance of the original MERCK isolate. Such an increase has been realized through a succession of at least 20 mutation-selection steps. In addition to these conventional but very efficient mutation-selection methods, powerful new techniques are recently available. Strains with different characteristics can undergo recombinaison by protoplast fusion (Gram-positive bacteria e.g. *Propionibacteria, Bacillus, Streptomyces)* or using conjugative plasmids (such as the IncP1 plasmids in Gram-negative bacteria e.g. *Pseudomonas*). The use of these techniques on the vitamin B12 producer organisms has not yet been reported in the literature. Genetic engineering of a system involving so many genes- the number and the chromosomal localization of which are still unknown- is a tremendous challenge which manufacturers are certain to accept.

REFERENCES

1. FUNK, C. (1911) J. Physio 43 : 395
2. GYÖRGY, P., WAGNER-JAUREGG, Th., KUHN, R (1933) Mitt. Kaiser. Wilhelm. Ges. 2, 1/4
3. SASAJIMA, K.I. ; YONEDA, M. (1971) Agr.Biol.Chem. 35 : 509
4. YOKOTA, A., SASAJIMA, K.I. (1981) Agr.Biol.Chem. 45 : 2417
5. French Patent No.6901867 (2001-108) assigned to Takeda Chemical Industries. (equivalent to U.S Patent No.3,607,648). (1969)
6. French Patent No.7438436 (2252-402) assigned to Takeda Chemical Industries. (equivalent to U.S Patent No.3,970,522) (1974)
7. DEMAIN, A.L. (1972) Ann. Rev. Microbiol. 11 : 369.
8. MINER, C.S. (1940) U.S Patent No.2,202,161 assigned to Commercial Solvents Corp.
9. MEADE, R.E., RODGERS, N.E., POLLARD, H.L. (1947) U.S. Patent No.2,433,063 assigned to Western Condensing Co.
10. LEVINE, H., OYAAS, J.E., WASSERMAN,L., HOOGERHEIDE,J.C., STERN R.M. (1949) Ind. Eng. Chem. 41 : 1665
11. GUILLIERMOND, A., FONTAINE, M., RAFFY,A. (1935) Compt. Rend. Acad.Sci.(Paris) 201 : 1077
12. MOSS, A.R., KLEIN,R. (1949) British Patent No.615,847 assigned to Roche Products.
13. HOST'ALEK,Z. (1957) J. Gen.Microbiol. 17 : 267
14. WICKERMAN,L.J., FLICKINGER,M.H., JOHNSTON,R.M. (1946) Arch. Biochem. 9 : 95
15. TANNER,F.W.Jr., VOJNOVICH,C., VAN LANEN,J.M. (1949) J. Bact. 58 : 737
16. MALZAHN,R.C., PHILLIPS,R.F., HANSON,A.M. (1959) U.S. Patent No.2,876,169 assigned to Grain Processing Corp.
17. CZCZESNIAK,T., KARABIN,L., WITUCH,K. (1973) Polish Patent No.66611 assigned to Instytut Przemyslu Farmaceutycznego-Warszawa
18. KAPLAN,L., DEMAIN,A.L. (1970) In Recent Trends in Yeast Research. D.G.Ahearn,Ed. p.137. Georgia State University.
19. MALZAHN,R.C., PHILLIPS,R.F., HANSON,A.M. (1963) Abst. 63rd. Annual Meeting. American Society for Microbiology p.21. Cleveland.
20. LAGO,B.D., KAPLAN,L. (1980) Adv. in Biotechnol. Vol III (Proceedings of the 6th. Intern. Ferment. Symp. London. Ontario)
21. BACHER,A., LINGENS,F. (1971) J.Biol.Chem. 246 : 7018
22. HOLLANDER,I.J,BROWN,G.M. (1979) Biochem. Biophys. Res. Commun. 89 : 759
23. HOLLANDER,I.J., BRAMAN,J.C.,BROWN,G.M. (1980) Biochem. Biophys. Res. Commun. 94 : 515
24. FOOR,F., BROWN,G.M. (1975) J.Biol.Chem. 250 : 3545
25. BURROWS,R.B., BROWN,G.M. (1978) J.Bact. 136 : 657
26. PLAUT,G.W.E. (1963) J.Biol. Chem. 238 : 2225
27. BACHER,A., BAUR,R., OLTMANNS,O., LINGENS,F. (1969) FEBS Letts. 5 : 316
28. OLTMANNS,O., BACHER,A., LINGENS,F., ZIMMERMANN,F.K. (1969) Mol. Gen. Genet. 105 : 306
29. BANDRIN,S.V., RABINOVICH, P.M., STEPANOV, A.I. (1983) Genetika 19 : 1419

30. BRESLER,S.E., CHEREPENKO,E.I., CHERNIK,T.P., KALININ,V.L., PERUMOV,D.A. (1970) Genetika 6 : 116
31. SHAVLOVSKII,G.M., KASHCENKO,V.E.,KOLTUN,L.V., LOGVINENKO,E.M., ZAKAL'SKII,A.E. (1976) Mikrobiol. 46 : 578
32. MITSUDA,H., SUZUKI,Y. (1970) J.Vitaminol. 16 : 172
33. SHAVLOVSKII,G.M., KOLTUN,L.V., KASHCENKO,V.E. (1978) Biokhimiya 43 : 2074
34. BRESLER,S.E., GLAZUNOV,E.A., GORINCHUK,G.F., CHERNIK,T.P., PERUMOV, D.A. (1978) Genetika 14 : 1530
35. BRESLER,S.E., GLAZUNOV,E.A., CHERNIK,T.P., SHEVCHENKO, T.N., PERUMOV, D.A. (1973) Genetika 9 : 84
36. GLAZUVOV,E.A., BRESLER,S.F., PERUMOV,D.A. (1974) Genetika 10 : 83
37. BRESLER,S.F., PERUMOV,D.A. (1979) Genetika 15 : 967
38. MINOT,G.R., MURPHY,W.P. (1926) J. Am. Med. Assoc. 87 : 470
39. RICKES,E.L., BRINK,N.G., KONIUSZY,F.R., WOOD,T.R., FOLKERS,K. (1948) Science 107 : 396
40. SMITH,E.L., PARKER,L.F.J. (1948) Chem. and Eng. News. 26 : 2218
41. HODGKIN, D.C. (1979). Vitamin B12. B. Zagalak, W. Friedrich,Edts. Walter de Gruyter and Co. Berlin. N.Y. p.19
42. LAMM,L., HECKMANN,G., RENZ,P. (1982) Eur. J. Biochem. 122 : 569
43. HÖLLRIEGL, V.,LAMM,L., ROWOLD,J., HÖRIG,J., RENZ,P. (1982) Arch.Microbiol. 132 : 155
44. FORD,S.H., FRIEDMANN,H.C. (1979) Biochim. Biophys. Acta. 569 : 153
45. BYKHOVSKI,V.Ya. (1979) Vitamin B12. B. Zagalak, W. Friedrich, Edts. Walter de Gruyter and Co. Berlin. N.Y. p. 293
46. HALL,H.H., BENEDICT,R.G., WIESEN,C.F., SMITH,C.E., JACKSON,R.W. (1953) Appl. Microbiol. 1 : 124
47. LEWIS,J.C., IJICHI,K., SNELL,N.S.,GARIBALDI,J.A. (1949) U.S. Department agriculture Bulletin AIC 254
48. PAGANO,J.F., GREENSPAN,G. (1954) U.S. Patent No.2,695,864 assigned to Olin Mathieson Chemical Corporation.
49. LONG,R.A., PARLING,N.J. (1962) U.S. Patent No.3,018,225 assigned to Merck and Co.
50. SPEEDIE,J.D., HULL,G.W. (1960) U.S. Patent No.2,951,017 assigned to Distiller's Company Limited.
51. DEMAIN,A.L., DANIELS,H.J., SCHNABLE,L., WHITE,R.F. (1968) Nature (London) 220 : 1324
52. WHITE,R.F., KAPLAN,L., BIRNBAUM,J. (1973) J.Bact. 113 : 218
53. WHITE,R.F., DEMAIN,A.L. (1971) Biochim. Biophys.Acta 237 : 112
54. KAPLAN,L., BIRNBAUM,J. (1973) Abst. Annual Meeting of the American Society for Microbiology. E.45.
55. KUCHERA,R.V., GEBGARDT,A.G. (1972) Prikl.Biokhim. Mikrobiol. 8 : 341

56. YONGSMITH,B., SONOMOTO,K., TANAKA,A., FUKUI,S. (1982) <u>Eur.</u> <u>J. Appl. Microbiol. Biotechnol.</u> 16 : 70
57. GRUZINA,V.D., EROKHINA,L.I., PONOMAREVA,G.M. (1973) <u>Genetika</u> 9 : 158
58. VOROB'EVA,L.I., BARANOVA,N.A., CHAN THI THAN, H. (1972) Primen. Khim. Mutagenov.Sel.Khoz.Med.Mater.Vses.Konf.Khim.Mutagenezu:103

INVERTASE ACTIVITY IN ENTRAPPED YEAST CELLS

G. Aykut, V.N. Hasirci, and N.G. Alaeddinoğlu

Dept.of Biological Sciences
Middle East Technical University
Ankara, Turkey

INTRODUCTION

The immobilization of intact microbial cells has recently attracted attention because of their potential for industrial applications. This interest stems mainly from the fact that im- mobilized whole cells favorably combine the advantages inherent in the use of immobilized enzymes with those of microbial fermentations. For instance enzyme extraction and purification are eliminated, higher yields of enzymes are obtained, retention of structural and confirmational integrity is achieved, greater potential is offered for multistep processes, and enzyme stability is increased.[6,7,8,12,25,26]

A large variety of immobilization methods are used for both enzyme and whole cell entrapment. These methods can be divided into chemical and physical methods. Immobilization by covalence is a chemical method whereas immobilization by adsorption, micro- encapsulation and lattice entrapment are physical methods.[3,6,8,19,20,21,22,27,29,30,31,32,33]

Immobilization by covalence is based on the formation of covalent bonds between the cell and the carrier. The coupling of the cell to the support can be achieved by either the activation of the support or the use of a coupling reagent to link the matrix and the cell. Supports such as carbodiimide-activated carboxymethyl cellulose and glutaraldehyde-activated amine glass beads are used in immobilization by covalence. This method of immobilization results in remarkably stable conjugate systems. The cells are not eluted by changes in pH, temperature and ionic strength. However, immobili- zation by covalence displays some disadvantages such as loss of enzyme activity due to either the method of coupling or the support material. Also many covalent binding techniques are comlicated and expensive. Although covalent methods are widely used for enzyme immobilization, they have not found as wide an application in whole cell immobilization. Thus, examples of covalently-linked micro-organisms are very few.[8]

The method of immobilization by adsorption consists of contacting an aqueous solution of a microorganism with a surface active adsorbant such as anion and cation exchange resins, cellulose and porous glass. Efficiency of adsorption is dependent on the characteristics of the adsorbant as well as the microorganism and experimental variables as pH, temperature and ionic strength.

Microencapsulation of whole cells may involve one of the following three methods:

    i) interfacial polymerization,
   ii) coacervation of preformed polymers, and
  iii) coacervation of liquid surfactant membranes.

In interfacial polymerization, an aqueous solution containing the cells and one component of the copolymer is emulsified in an organic solvent with a soluble organic surfactant. This emulsion is later mixed with a membrane forming reagent dissolved in the same organic solvent.

In the coacervation dependent process, the aqueous solution contains the cells only, and the polymer that forms the semipermeable membrane is in the added solvent.

In the liquid surfactant membrane process, the cell suspension is added, in drops, into a cooled water-immiscible oil phase containing a surfactant.

The final method involves the entrapment of cells within the interstitial spaces of water insoluble polymers. The method aims at the formation of a highly cross-linked network of a polymer in the presence of whole cells. Thus the cells are physically entrapped within the polymer lattice. The homogeneous dispersion of whole cells results in greater enzyme accesibility to the substrate and faster reaction rates. Leakage of cells from the gel matrix can be overcome by effective cross-linking. The chemicals necessary for polymerization should be used with caution since they have the potential to damage the enzyme or the microorganism. Almost all of the above methods have been tried in the immobilization of Saccharomyces cerevisiae.[1,4,5,10,12,14,17,18]

The present study discusses the feasibility of immobilizing β-fructofuronosidase(E.C.3.2.1.26)for the conversion of sucrose into glucose and fructose by polyacrylamide gel-entraped whole cells of

S.cerevisiae. When the potential of fructose as a natural sweetener is considered, the importance of this convertion becomes obvious. The commonly practised method of acid hydrolysis of sucrose imposes the limitation of low yields. Enzymatic hydrolysis, on the other hand, provides high conversion rates and can overcome undesirable effects such as discoloration.[4] The technique has been further improved by immobilization of isolated and purified enzyme from a variety of sources, as well as immobilization of whole cells.[4,5,24]

For our purpose, cross-linked polyacrylamide was chosen since it promised to fulfill the following requirements: i) no chemical modification ii) easy preparation iii) good flexibility with variable sizes of the lattice formed.

Initially, characterization of gels, with varying total concentrations and ratios of monomer and cross-linking agent, was performed. Then the optimal conditions for sucrose hydrolysis with both free and immobilized yeast cells were determined.

MATERIAL and METHODS

Preparation of Gels

Gels were prepared by dissolving acrylamide, bisacrylamide and ammonium persulfate in different ratios (Table 1) in 5 ml of isotonic phosphate buffer (pH 7.6) and incubating the resulting solution in an uncovered glass tube of fixed dimension, at 37°C for 90 min.[16] After synthesis, the gels were recovered as a slab.

In gels Nos.1,2, and 3, the ratio of acrylamide to bisacrylamide was varied with polimerization mixture concentration remaining constant; in gels Nos.1,4, and 5, the concentration was varied keeping the acrylamide to bisacrylamide ratio constant.

Each gel was swollen in water for 19 days, in ethanol for 19 days and in chloroform for 10 days. The weights of the swollen gels were determined at varying intervals and the dry weights of the gels were obtained by drying in a vacuum oven at 50°C.

Solvent content (W%), swelling ratio (Q) and the volume of adsorbed solvent (V.A.S) for each gel in solvents of varying polarity was calculated according to the following equation;[9]

$$W\% = \frac{W_s - W_d}{W_d} \times 100 \; ; \; Q = \frac{W_s}{W_d} \; ; \; V.A.S = \frac{(W_s - W_d)/q_3}{W_d}$$

Where W_s : weight of swollen gel

W_d : weight of dry gel

q_3 : density of solvent at room temperature

Table 1. Input concentrations of acrylamide (A), N, N'-methylenebis-acrylamide (BAA) and ammonium persulfate in the synthesis of polyacrylamide gels.

Gel Number	Acrylamide (g/5 cm^3)	BIS (g/5 cm^3)	Ammonium Persulfate (g/5 cm^3)	% Polymer (W/V)	(A)/(BAA)*
1	0.75	0.04	0.0125	15.8	40.6
2	0.75	0.08	0.0125	16.6	20.2
3	0.75	0.02	0.0125	15.4	82.3
4	1.50	0.08	0.025	31.6	40.6
5	0.375	0.02	0.00625	7.9	40.6

*Molar ratio.

Preparation of Cells and Growth Media

The yeast used in gel entrapment was isolated from bread dough and identified as a strain of S.cerevisiae. The cells were grown aerobically in a growth medium containing N.Broth (Oxoid), 1% sodium lactate and 0.2% glucose and incubated at 30°C for 48h. After growth, cells were centrifuged (at 6600g) and resuspended in cold isotonic phosphate buffer. The appropriate amounts of this cell suspension were used as source of invertase for free and immobilized cells. The amount of cells used is given as the wet weight of the packed cells.

Determination of Invertase Activity

Of the several methods available, the Nelson method[28] was selected. The method makes use of the reaction in which glucose and fructose reduce the Cu^{+2} complex to $Cu_2O.Cu_2O$, which then quantitatively converts arsenomolybdic acid (yellow) to arsenomolybdous acid(blue) which can be determined colorimetrically at 540nm. Assay reaction time was 30 min.and the reaction volume was 7 ml. Both for free and immobilized cells, units of enzyme activity are defined as mmol glucose/min.

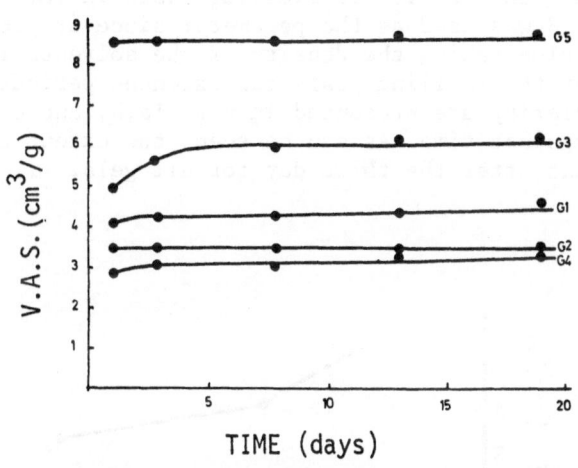

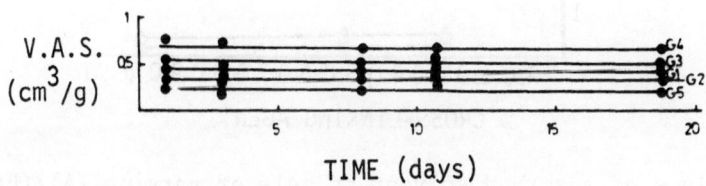

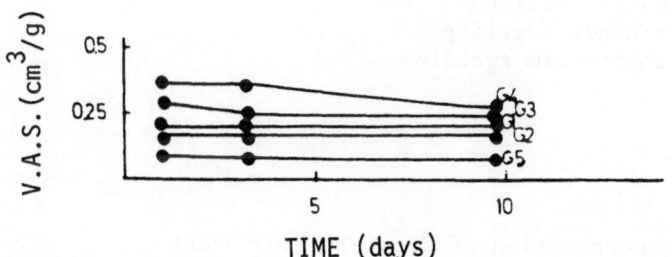

Fig. 1(a,b,and c); Equilibration of gels Nos. 1,2,3,4, and 5 with water, ethanol and chloroform in time.

175

RESULTS

Swelling Tests

To compare the results of swelling tests in various organic
solvents, V.A.S was used as the parameter since it yielded the most
appropriate value taking the density of the solvents into account.
The results of the swelling tests for extended periods in solvents
of varying polarity are presented in Fig. 1a,b, and c where V.A.S.
was plotted against time. As can be seen, the extent of swelling is
almost constant after the third day for all gels.

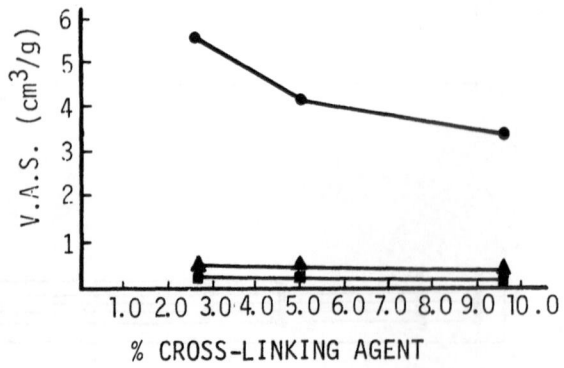

Fig. 2.Volume of adsorbed solvent by gels of varying (A)/(BAA) ratio.

●: water swelling
▲: ethanol swelling
■: chloroform swelling

Effect of Concentration of Cross-linking Agent

Using the results from the third day of swelling, V.A.S. values
for gels Nos.1,2, and 3 and for gels Nos.1,4,and 5 are shown in
Figs.2and 3 respectively. For gel Nos.1,2 and 3, where the
polimerization mixture concentration was kept constant and the
acrylamide to bisacrylamide ratio varied, the effect of an increase
in the content of cross-linking agent is to a decrease the volume
of adsorbed water. A similar behaviour was observed for the volume
of adsorbed ethanol and chloroform.

Effect of Concentration Polymerization Mixture on Swelling

For gels Nos.1,4,annd 5, where the acrylamide to bisacrylamide
ratio was kept constant and the polimerization mixture concentration
varied, the volume of adsorbed water decreased with increasing
concentration; the volume of adsorbed ethanol and chloroform increased
only slightly. Thus the pattern followed by the less polar solvents
differs from that of water for these gels (Fig.3).

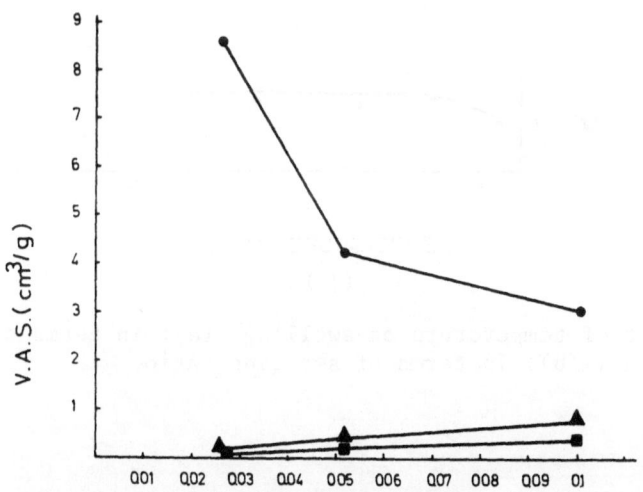

POLYMERIZATION MIXTURE CONCENTRATION (M)

Fig.3.Volume of adsorbed solvent by gels of varying mixture
concentrations.

●:water swelling
▲:ethanol swelling
■:chloroform swelling

Effect of Temperature and pH on Swelling

Using gel No.1 with 16% (w/v) polymer and a 0.976/0.024 acrylamide to bisacrylamide molar ratio, the temperature effect on
swelling content (W%) and swelling ratio(Q) was determined (Fig 4a,
and b). Increases in the temperature of the medium from 4°C to 20°C
increased the volume of adsorbed solvent indicating that swelling
of the gel, in that interval, is endothermic. Swelling content and
swelling ratio determined at different pH values indicated that
changes in pH did not affect the swelling ability of the gel.

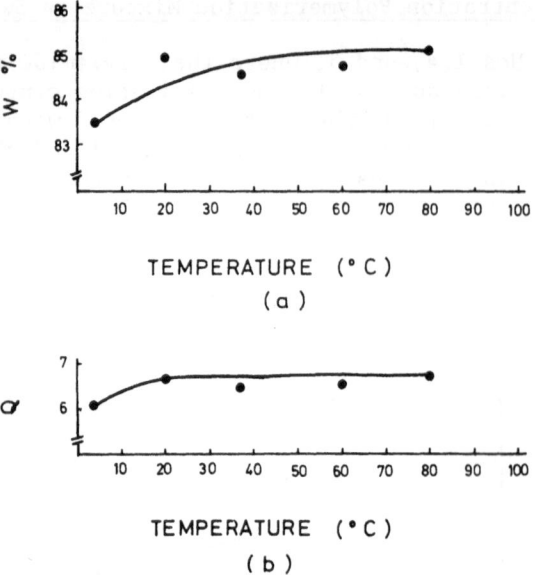

Fig. 4.Effect of temperature on swelling; (a): in terms of swelling
content, (b): in terms of swelling ratio.

ENZYMATIC PROPERTIES OF FREE AND IMMOBILIZED CELLS

The responses of free cells to parameters such as substrate
concentration, pH and temperature usually differ from those of the
immobilized cells. The difference resides in the physical and chemical
properties of the carriers that alter the microenvironment of the
bound cells. Therefore each parameter has to be resolved for both
the systems.

Effect of Substrate Concentration with Free Cells

The reaction was carried out at pH 4.0 at 25°C in the range
32-940 mM sucrose concentration. As shown in Fig.5, invertase
activity was inhibited by sucrose concentrations higher than 350 mM.
No considerable effect of substrate concentration was observed in
the range 32 -313 mM.

Effect of pH on Free Cells

Reactions were carried out at 25°C, with 200 mM sucrose and at
various pH values (Fig.6). Optimal invertase activity was exhibited
at pH 3.6.

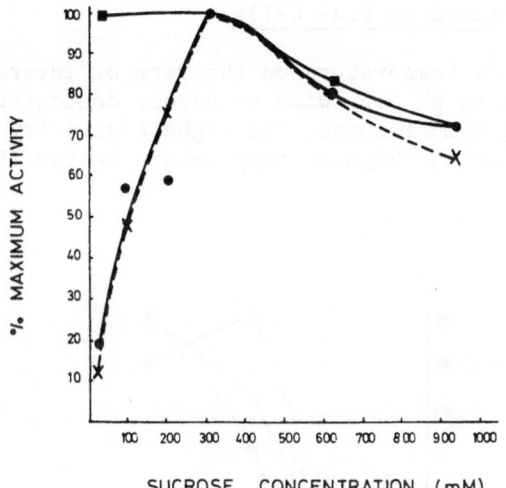

SUCROSE CONCENTRATION (mM)

Fig. 5. Effect of substrate concentration on invertase activity
(■-■:free cells ; ●-●: 1% cell loaded gel ; x-x: 20% cell
loaded gel).

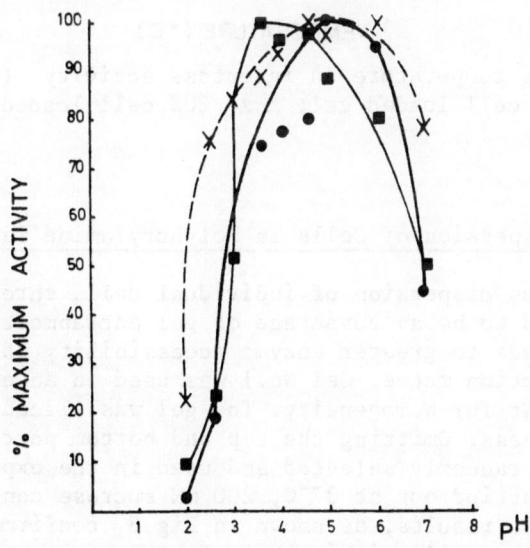

Fig. 6. Effect of pH on invertase activity (■-■: free cells; ●-●:
1% cell loaded gel; x-x: 20% cell loaded gel).

179

Effect of Temperature on Free Cells

The effect of temperature on the rate of inversion of sucrose was investigated at pH 3.6, with a sucrose concentration of 200 mM, and with varying temperatures. The highest activity of free cells was found to be at the highest temperature tested (Fig.7).

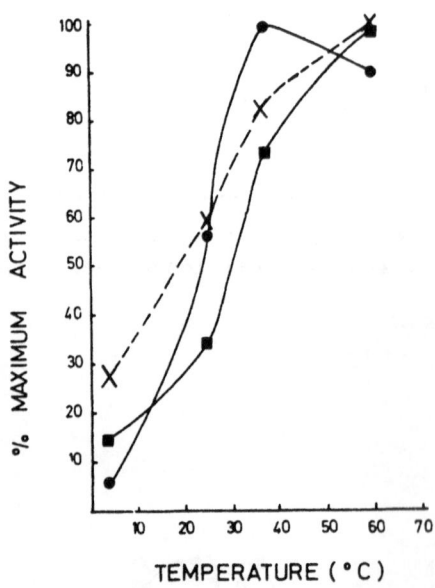

Fig.7. Effect of temperature on invertase activity (■-■: free cells; ●-● : 1% cell loaded gel; x-x: 20% cell loaded gel).

Homogeneous Dispersion of Cells in Polyacrylamide Gel

Homogeneous dispersion of individual cells throughout the gel has been stated to be an advantage of gel entrapment since the homogeneity leads to greater enzyme accessibility to the substrate and faster reaction rates. Gel No.1 was used in an experiment designed to test for homogeneity. The gel was sliced into 30 pellets of equal thickness. Omitting the top and bottom portions, 8 pellets out of 30 were randomly selected and used in the experiment. The reaction was carried out at $37^{\circ}C$, 200 mM sucrose concentration and a pH of 3.6. The results, as shown in Fig.8, confirmed a homogeneous dispersion of the cells within the gels tested.

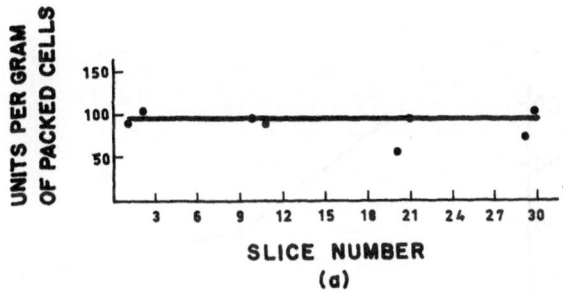

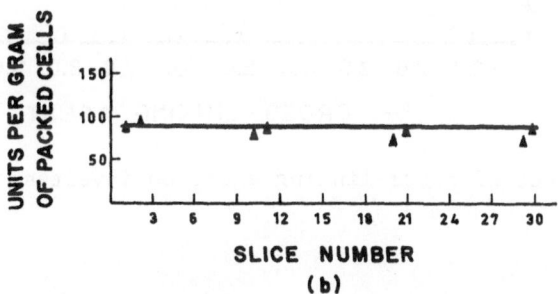

Fig. 8. Dispersion of S.cerevisiae cells in gel No.1 (a:trial 1; b : trial 2).

Effect of Gel Structure

As found earlier, with increasing cross-linking agent content, swelling in water decreases; similarly swelling in water decreases in gels prepared in concentrated solutions. In selecting one of the five gels to be studied, one would anticipate the invertase activity to increase with increased swelling. However, the gel that produced the maximum invertase activity would also be the one that leaked most. Taking this into consideration, cells were entrapped in each gel and the activities were compared with the amount of leakage. Invertase activity produced by gels Nos.1,2 and 3 (where polymerization mixture concentration was kept constant and the acrylamide to bisacrylamide ratio was varied) is shown in Fig.9. Invertase activities with gels Nos.1,4 and 5 (where the acrylamide to bisacrylamide ratio was kept constant and concentration of polymerization mixture was varied) is represented in Fig.10.

The results of leakage experiments (even for extended periods of soaking) indicated that there was practically no leakage from any of the gel structures, tested. These findings led us to choose gel No.5 for which the relative molar ratio of acrylamide to bisacrylamide as 0.976/0.024 with 7.9% polymer concentration.

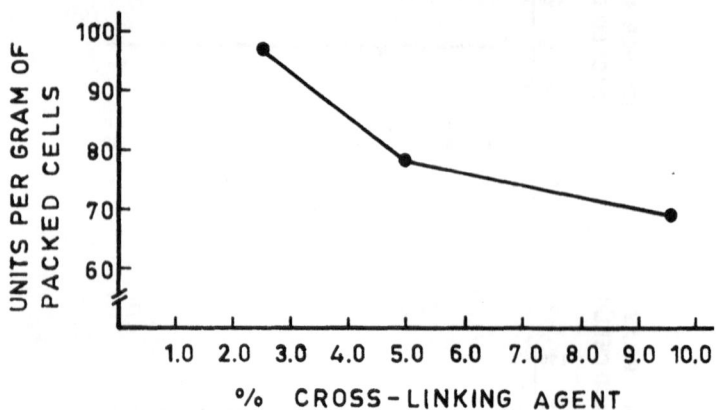

Fig. 9. Effect of cross-linking agent on invertase activity.

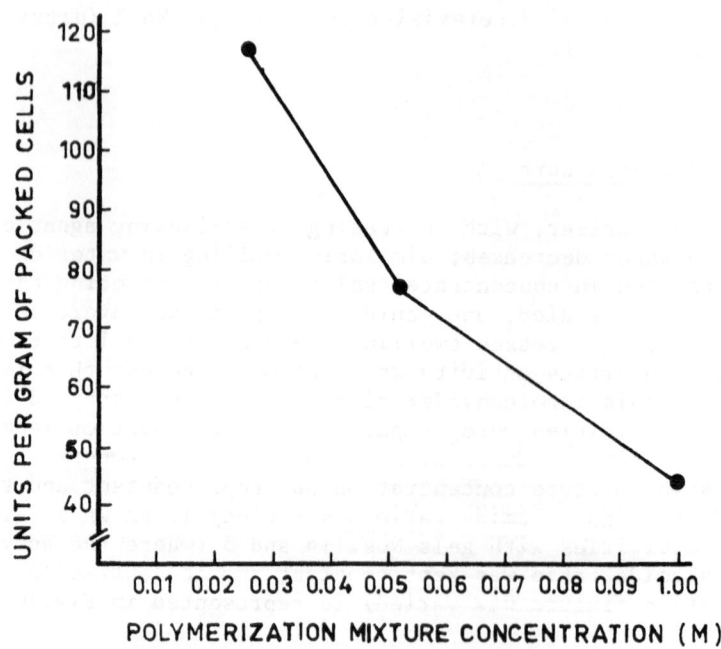

Fig. 10. Effect of mixture concentration on invertase activity.

Effect of Cell Loading

Gels prepared with different concentrations of cells displayed maximum specific activity and maximum absolute activity with 1% and 20% cell loading respectively (Fig.11, a and b).

Fall in the specific activity with immobilization of more concentrated cell suspensions or solutions have been reported. Durand and Navarro (1978) state that when more concentrated cell suspensions were entrapped, the efficiencey, with which they carried out a particular enzymic activity fell to very low levels. Similarly, Onyezili and Onitiri (1981) reported that the specific activity of invertase immobilized on O-alkylated nylon tubes decreased with increasing concentration of enzyme used in coupling step.

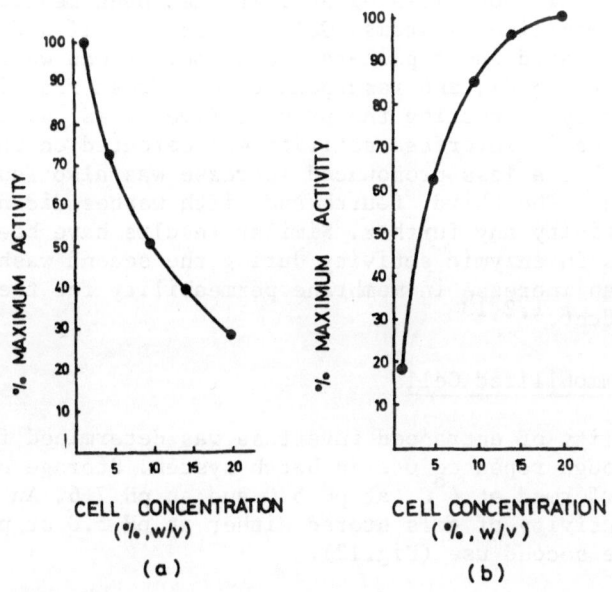

Fig. 11. Effect of cell loading on immobilized invertase activity. In (a) activity is the specific activity-units per gram of packed cells; in (b) activity is the total activity-units per 5 cm^3 gel.

Effect of Substrate Concentration with Immobilized Cells

Gel pellets, with 1% and 20% cell content respectively, were assayed at 37°C, pH 3.6 with varying sucrose concentrations ranging from 32-940 mM. The invertase activity of both the gels was inhibited at sucrose concentrations greater than 350 mM (Fig.5).

Effect of pH

The reaction was held at 37°C at a 200 mM sucrose concentration. Both gels exhibited a pH optimum of 5.0-6.0 (Fig.6).

Effect of Temperature

Gel pellets, with 1% and 20% cell load, were assayed at various temperatures with a sucrose concentration of 200 mM. The entrapped enzyme was maximally active at 30-40°C when the gel was 1% cell loaded and it was maximally active at the highest temperature (60°C) tested (Fig.7).

Reuse of Immobilized Cells

Repeated use without loss of activity has been mentioned as an advantage of immobilized systems. Gel pellets with 1% and 20% cell loading were prepared in triplicates; each pellet was washed with acetate buffer (pH 5.0), and resuspended in a fresh reaction mixture, assayed separately, repeating the process five times. With a 1% cell load, an increase in invertase activity was detected on the second use. A similar but a less pronounced increase was also found with 20% cell loading. The third, fourth and fifth washes did not alter the enzymic activity any further. Similar results have been found and the changes in enzymic activity during the second wash have been attributed to an increase in membrane permeability for the substrate and/or the product.[1,2,13]

Stability of Immobilized Cells

The stability of entrapped invertase was determined for a period of 18 days through repeated use in batch system. Storage of gel pellets was performed at 4°C, at pH 5.0 and at pH 7.6. An increase in invertase activity of gels stored either at pH 5.0 or pH 7.6 was detected on the second use (Fig.12).

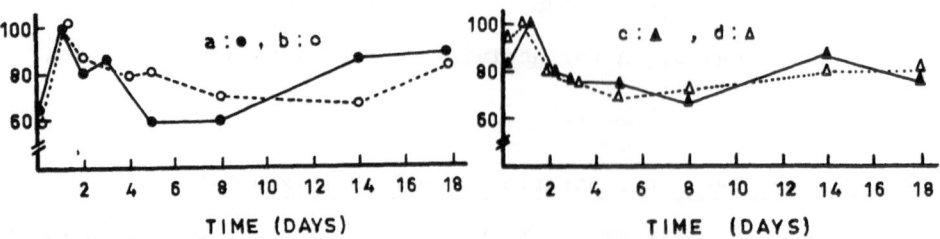

Fig. 12. Stability of immobilized invertase activity. (a): 1% cell loaded gel stored at pH 7.6; (b): 1% cell loaded gel stored at pH 5.0; (c): 20% cell loaded gel stored at pH 7.6; (d): 20% cell loaded gel stored at pH 5.0.

DISCUSSION

Characterization of the systems used as supports in immobilization is not a common practice. Even though, in most cases, suitable conditions for the immobilization and the enzymic properties of the cells have been determined, no information on the physical properties of the immobilizing support have been given. In the present work, characterization of polyacrylamide gels have been accomplished before the entrapment of cells. Later in the work, the trend of invertase activity with immobilized cells, in going from the least cross-linked to the most cross-linked gel and from the least concentrated to the most concentrated gel was determined.

In comparing the results obtained with free and entrapped cells, the extent of inhibition by the substrate was slightly higher for the entrapped system. This can be attributed to a decrease in the rate of diffusion of substrate to the enzyme due to increased viscosity of the sucrose solution.

Free enzyme exhibited an optimum pH at 3.6 and the entrapped enzyme at pH 5.0 -6.0. Alterations in pH profiles of enzymes immobilized on charged carriers are believed to occur due to a difference in the electrostatic, field around the immobilized enzyme caused by the inherent charge of the carrier.[23] However, it is not possible to attribute the pH shift in our case to the charge of the matrix since the water-insoluble matrix formed from polyacrylamide is electrically neutral.[32] No pH shift has been reported for invertase of poly-acrylamide entrapped S.cerevisiae cells.[4] However a similar unexpected shift towards more acid pH's has been described for phosphoglycerate mutase[23,32] and aspartase of Escherichia coli entrapped in polyacrylamide gel. This unexpected behaviour can be[23] explained by alterations in the environment within the gel matrix or by alterations in the cell due to destruction of amino and carboxyl groups on immobilization.[13] The results suggesting that immobilization increased the pH stability of the enzyme are in good agreement with literature reports.[4,11,23] Under similar optimum conditions D'Souza (1980) reported a 15% activity loss on immobilization of S.cerevisiae in rapidiopolymerized polyacrylamide gels. Our results (data not shown) revealed about 40% activity loss with 1% cell loading and 80% activity loss with 20% cell loading respectively. The loss of activity may be attributed to inactivation of the enzyme during the prolonged periods required for chemical polymerization of acrylamide.

REFERENCES

1. Chibata, I., Tosa, T.. Sato, T. Immobilized Aspartase containing microbial cells: Preparation and Enzymatic Properties. App. Microbiol. 27(5): 878-885 (1974).

2. Chibata, I., Tosa, T., Sato, T. Production of L-aspartic acid by microbial cells entrapped in polyacrylamide gels. Methods in Enzymology, Vol.44, Academic Press Inc., New York (1976).
3. Chibata, I. Immobilized microbial cells with polyacrylamide gel and carrageenan and their industrial applications, Chapter 13 in immobilized microbial cells, Venkatsubramanian, K.Ed., ACS Symposium Series 106, ACS, Washington, D.C (1979).
4. D'Souza, S.F., Nadkarni, G.B. Continious inversion of sucrose by gel entrapped yeast cells. Enzyme Microb. Technol. 2: 217-222 (1980).
5. D'Souza, S.F., Nadkarni, G.B. Continious conversion of sucrose to fructose and gluconic acid by immobilized yeast cell multienzyme complex. Biotechnol. Bioeng. 22 : 2179-2189 (1980)
6. Durand, G., Navarro, J.M. Immobilized microbial cells. Process Biochem. 13(9) : 14-23 (1978).
7. Evrimler, M., Sonaer, H., Çağlar, A. Poliakrilamid Jelinde tutuklanmış Acetobacter suboxydans. Ulusal Biyomühendislik Kongre Tebliğleri, Lider Matbaacılık, Ankara (1981).
8. Fukui, S., Tanaka, A. Immobilized microbial cells. Ann.Rev. Microbiol. 36 : 145-172 (1982).
9. Hasırcı, V.N. Synthesis and characterization of PVNO and PVNO-PVP hydrogels. Biomaterials. 2 : 8-12 (1981).
10. Isaeve, V.S., Kolpakchi, A.P. Fixation of brewer's yeast to polymer materials. Prikl. Biokhim Mikrobiol. 12(6) : 866-870 (1976).
11. Kawashima, K., Umeda, K. Immobilization of enzymes by radiopolymerization of acrylamide. Biotechnol. Bioeng. 16 : 609-621 (1974).
12. Kennedy, J.F., Barker, S.A., Humphreys, J.D. Microbial cells living immobilized on metal hydroxides. Nature. 261 : 242-244 (1976).
13. Kimura, A. Research Institute for Food Science, Kyoto Univ., Uji, Kyoto 611, Japan. Personal Communication (1983).
14. Navarro, J.M. Fermentation en continue a l'aide de microorganismes fixés. Thesis Doct. Ing. Univ. Toulouse (1975).
15. Onyezili, F.N., Onitiri, A.C. Immobilization of invertase on modified nylon tubes. Anal. Biochem 113 : 203-206 (1981)
16. Rembaum, S.P., Yen, S.P.S., Ingram, M., Newton, J.F., HU, C.L., Frasher, G.W., Barbour, B.H. Platelet adhesion to heparin-bonded and heparin-free surfaces. Biomat., Med.Dev., Artificial Organs 1(1) : 99-119 (1973)
17. Samejima, H., Kimura, K., Ado, Y., Suzuki, Y., Tadokoro, T. Regeneration of ATP by immobilized microbial cells and its utilization for synthesis of nucleotides. Enzyme Eng. 4 : 237-244 (1978).
18. Sato, T., Nishida, Y., Tasa, T., Chibata, I. Immobilization of Escherichia coli cells containing aspartase activity with K-carrageenan. Biochem. Biophys. Acta. 570 : 179-186 (1979).

19. Sidney, J.G. Physical Techniques : Entrapment in immobilized enzyme preparation and engineering techniques, Noyes Data Corp. New Jersey (1974).
20. Sidney, C.P., Kaplan, N.O. Covalent coupling in Methods in Enzymology (Section II.A), Vol.44,Mosbach, K.Ed., Academic Press Inc., New York (1976).
21. Sidney, C.P., Kaplan, N.O. Adsorption in Methods in Enzymology (section II, B), Vol.44,Mosbach, K.Ed., Academic Press Inc., New York (1976).
22. Sidney, C.P., Kaplan, N.O. Entrapment and related techniques in Methods in Enzymology (section II.C), Vol.44, Mosbach, K.Ed., Academic Press, Inc., New York (1976).
23. Smiley, K.L., Strandberg, G.W. Immobilized enzymes, Weast, R.C. Ed., CRC Press, Ohio, 13-38 (1973).
24. Toda, K., Shoda, M. Sucrose inversion by immobilized yeast cells in a complete mixing reactor, Biotechnol. Bioeng. 17 : 481-497 (1975).
25. Vandamme, E.J. Immobilized microbial cells as catalysts, Chem. and Ind. 24 : 1070-1072 (1976).
26. Vieth, W.R., Venkatsubramanian, K. Immobilized microbial cells in complex biocatalysis, Chapter I in Immobilized Microbial Cells, Venkatsubramanian, K.Ed., ACS Symposium Series 106, ACS, Washington, D.C. (1979).
27. Wayne, P.H., Jr. Introduction to Immobilized enzymes. Chapter I in Immobilized Enzymes for Food Processing, Wayne, P.H., Jr. Ed., CRS Press, Inc. Florida (1980).
28. Wharton, D.C and McCarty, R.E. Experiments in Biochemistry. p.313, The Macmillan Company, New York (1972).
29. Zaborsky, O. Covalent attachment to water-insoluble functionalized polymers. Chapter 2 in Immobilized Enzymes, Weast, R.C. Ed., CRS Press, Ohio (1973).
30. Zaborsky, O. Properties of covalently bonded water-insoluble enzyme-polymer conjugates. Chapter 2 in Immobilized Enzymes. Weast, R.C. Ed., CRC Press, Ohio (1973)
31. Zaborsky, O. Adsorption. Chapter 5 in Immobilized Enzymes, Weast, R.C.Ed., CRC Press, Ohio (1973).
32. Zaborsky, O., Entrapment within cross-linked polymers. Chapter 6 in Immobilized Enzymes, Weast, R.C.Ed., CRC Press, Ohio (1973).
33. Zaborsky, O. Microencapsulation. Chapter 7 in Immobilized Enzymes, Weast, R.C.Ed., CRC Press, Ohio (1973).

COMPUTER-CONTROL OF FERMENTATION PROCESSES

D.R. De Buyser, J.A. Spriet[*] and G.C. Vansteenkiste

Seminar for Applied Mathematics and Biometrics
Coupure Links 653
9000 Gent, Belgium

INTRODUCTION

Already since the 18th century one can witness a steady trend towards the use of machines to replace the muscle power of animals and man. The rise of the digital computer in the fifties has speeded up this process. In the beginning, the computer was used for lower level mental processes such as information retrieval and tedious calculations. Nowadays the digital machine can replace higher level mental processes such as the ability to recognize patterns, to identify systems, to make decisions and so on.

In the last decade computerization of the human world has emerged in different production processes. Especially in the fields of oil refining, metallurgy, as well as in the chemical industry, automatization by computer is wide-spread and even common place. The cost of the computer-hardware decreases also constantly so that the computer can be introduced in the laboratory. More and more analytical instrumentation and laboratory fermentors, which can be interfaced to computer-hardware are available on the market now. The production-plant however has remained traditional for a long time in the fermentation world. The main reasons are (1) a lack of adequate sensors for product, substrate and biomass ; (2) a shortness of reliable process-models for process-control and (3) investments for a computer are comparatively larger in the fermentation field than in other industrial fields because of the smaller production scale.

---

[*] Belgian Fund for Scientific Research

In view of these considerations, the presentation will focus on different aspects of computer controlled fermentation. First, the hardware with its issues is introduced. Second, some typical problems encountered in computer-control of microbial processes and possible solutions are discussed. Third, the software configuration, especially for on-line process analysis and process control, is described. Finally, a conclusion and the future developments which may contribute to the control of fermentations, is brought up.

HARDWARE

It is not the intention here, to specify completely fermentation and computer-hardware. It is assumed that the reader has a fair understanding of the units mentioned. Attention mainly will be given to a few delicate points : (1) the fundamental components necessary to interface a computer with a fermentor ; (2) the controversy of using direct digital control or set-point control.

Some fundamental components

The fermentor
At present the culture apparatus is often designed to meet the requirements of a certain process. Nevertheless an often occurring fermentor type can be described as a working example. Fermentation equipment normally comprises a stainless steel or a glass vessel with input and output ports, an agitation set with motor drive, shaft and impeller, piping for providing air to the vessel and a heating system capable of maintaining sterilization and operational conditions.

A stumbling block for computerization, nowadays, is still the philosophy in industrial practice of constructing fermentors. The whole approach towards fermentor design is still focused on manual operation. If computer-coupling is explicitly mentioned by the client, the manufacturer will provide sensors with electrical outputs and eventually control inputs that can be set by the computer. Unfortunately this is only a first step towards full automation of the complete fermentor equipment. A better design would permit much greater communication facilities between fermentor and computer so that all sequencing and control can be set under computer supervision and that the 'state' of the fermentor can be provided to the computer in much more detail.
Photo 1 represents a computer-coupled fermentor.

The computer
Photo 2 is an illustration of a computer-system. The main elements of the digital machine are its memory that can store numerical values and text, and an arithmetic unit that can perform calculations, checks and logical operations. The actual activity of the computer is determined by a set of programs, or in general by its 'software'. To be useful to man, the machine has some form of communication. Therefore, there are some input-facilities like typewriters and keyboards and output-facilities like displays, printers and plotters.

190

Photo 1 : Fermentor with computer-interface and process-operator console

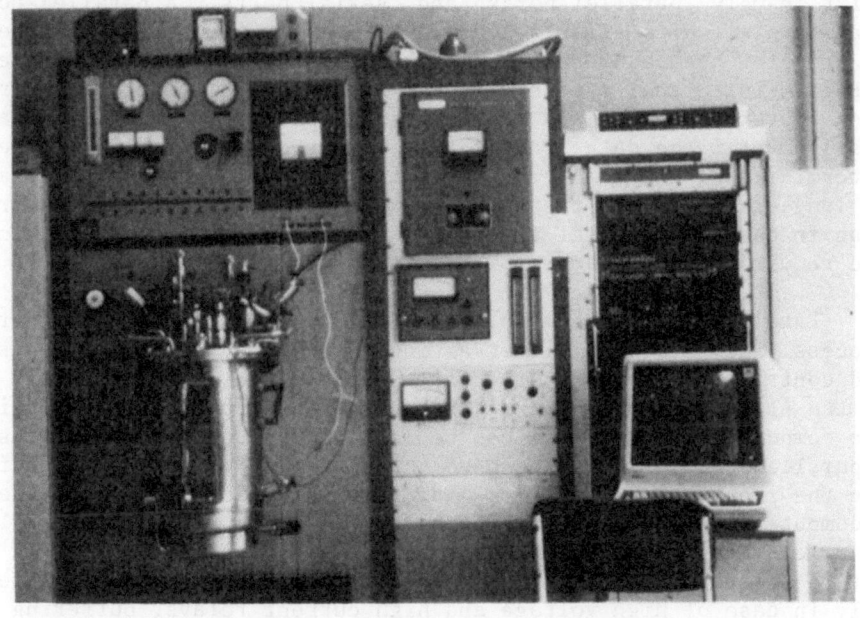

Photo 2 : Computer-configuration

191

Also some store-facilities are available, such as tapes and disks. To communicate with the process there is some form of interface, treated in the next section.

On the whole there are no stringent hardware requirements for the computer. The digital machine should be suited for process control and thus have an internal clock and the necessary hardware provisions for communication with the process and the operator. If only data-logging and simple control actions are required, the speed of the machine may be relatively low as the time constants of conventional fermentation processes are large. In view of those remarks, sometimes one minicomputer is utilized to supervise a whole battery of fermentors. However, the automated control of fermentation processes is not without its pecularities as will be shown in the sequel. As soon as more advanced activities like estimation, modelling and time optimal control of the process itself is considered the computational load on the machine increases substantially and small machines will be too slow. In those cases one should care to obtain substantial computing power and sufficient data storage.

Fermentor-computer interface

At the computer-side input and output ports are used to transfer data and/or control signals to and from the process. The input port is used to transfer the signals into the computer while the output port is used to transfer the signals from the computer to the fermentation system. Two different kinds of input-output ports exist, namely 'parallel ports' and 'serial ports'. A parallel port transfers all bits simultaneously while a serial port is designed to transfer the data one bit at a time. Normally serial links (e.g. a RS232C-interface) are used for communication between e.g. a process operator console and the process computer. A serial link can easily master the data traffic that exists on a connection between a terminal and the computer. In addition, if the console is located at long distance from the digital machine fewer links are required than in case of parallel transmission where at least as many lines are required as the number of bits transferred.

Parallel ports are mainly used in direct conjunction with the process. They are used either as input and output ports for sense and control lines or as ports for the input and output of data. If the state of a push-button or a switch has to be introduced in the computer, a connection can be made immediately with the bits of a parallel port without the need for an expensive interface. Buffering the input however ensures that the system does not crash when a computer overload occurs. In the same manner, as soon as a valve or a relay or some other on/off control has to be handled, only one bit of a parallel port is in principle required. Here also, especially in case of high voltage and high current relays, buffering is necessary. For analogue signals additional hardware is required. If the computer wants to collect data from a sensor through a meter,

an Analogue Digital Converter is used. This device converts the analogue quantities into digital ones. One can distinguish 8 bit, 12 bit
and 16 bit ADC's on the market. In view of the necessary precision
of the obtained figures, it is advisable to use at least 12 bit ADC's.
In practice, this means for instance for a 12 bit ADC a conversion
of the range -10 V to +10 V into an integer in the range of -2048
to 2048, which is acceptable.

As analogue digital converters are rather expensive, usually
one integrates a multiplexer with the ADC and some "sample and hold"
device. A multiplexer is an instrument that selects a certain sensor
or a certain control signal. The "sample and hold" device is a circuit that holds the instanteneous value of a signal. In this way a
great number of sensor-outputs can be converted by just one ADC.
In a fermentation-context, this should give no problems in accuracy
and in timing because the sample frequency lies always in the order
of minutes while the conversion rate of an ADC is often below the
millisecond. The software however must be properly adapted to the
hardware requirements.

For control purposes often the converse has to be performed :
a digital number has to be converted to an analogue signal in order
to feed the controller properly. Here a Digital to Analogue Converter (DAC) is used. This device performs the opposite conversion of
the ADC. One can distinguish again 8 bit-, 12 bit- and 16 bit DAC's.
Also here it is advisable to use at least 12 bit DAC's. As DAC's
are not so expensive as ADC's, DAC's are usually not multiplexed.

As presented up to here quite a number of parallel ports would
be required to access the fermentor completely. Because time is not
especially critical in a fermentation context, quite often only one
parallel port will be provided. Then addressing capability is required in order to properly single out that part of the fermentor
that is accessed at a given time. Software then takes care of proper
addressing and proper data transmission. It is clear that such an
approach requires more buffering as the separate units in the fermentor must be able to stand alone as long as they are not connected
directly with the single parallel port.
Figure 1 gives a control-loop for controlling the speed of the agi-
tator. The loop is connected to the parallel port of a minicomputer.
One very important remark in this loop can be made, namely it is
good practice to separate galvanically measuring device and computer-
system by the use of opto-couplers or action-packs, as represented
in figure 1.

With these optical isolators, the grounding problems in the
ditital signals are removed and moreover the transmission is by a
current loop which enhances noise rejection considerably. From own
experience and also stated by others (ROLF et al., 1982) this is
of importance in interfacing instruments with a floating reference,
such as pH-meters.

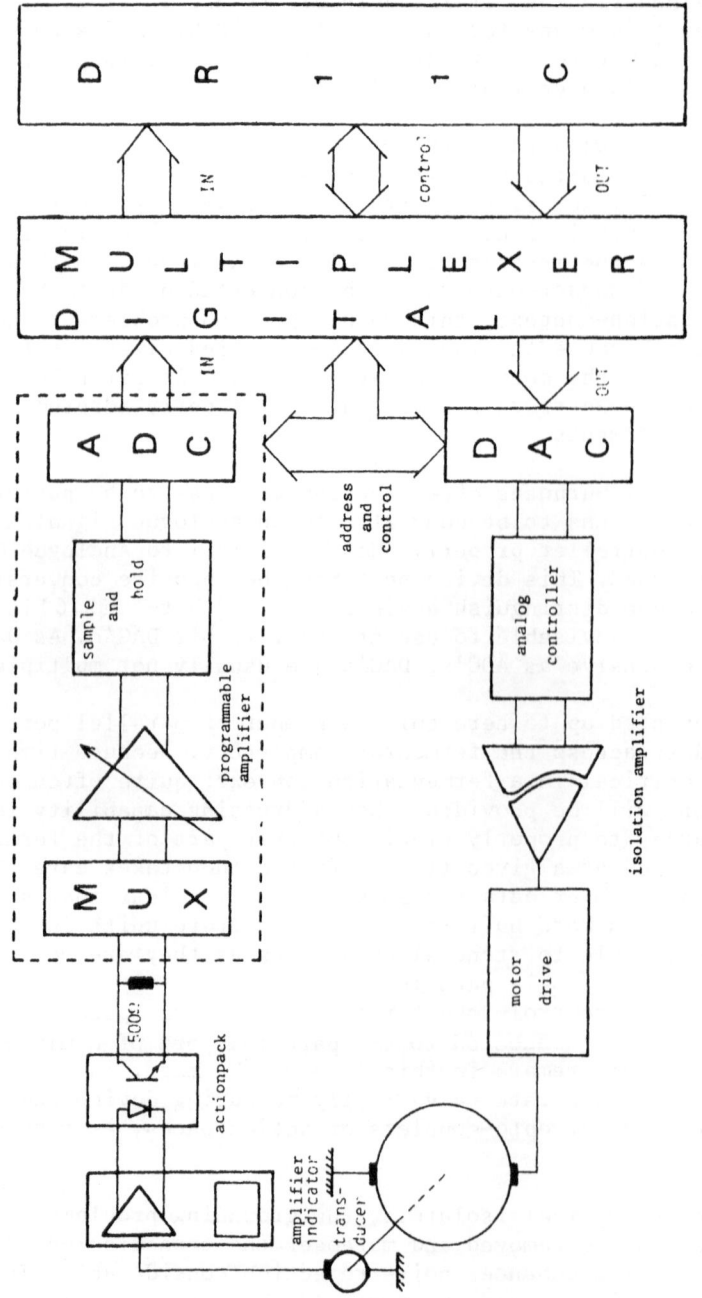

Figure 1 : Control-loop for agitation-control

As a conclusion of this section, one can state that normally no problems occur for reading the different sensors, for they can be provided with analogue electrical outputs. However, it may still be difficult to find an analogue controller, which setpoint can be set by an electrical signal. Also it may be a good advice that the customer gathers precise detail on measurement ranges of the sensors and their installation requirements and explicitly checks those items for adequateness.

Direct digital control versus set-point control

In a non-computerized plant, control actions are only taken by analogue controllers. The process instrumentation is equipped with analogue regulatory mechanisms that keep the controlled variable at a given setpoint. This setpoint is chosen manually. If the control profiles are not constant, the operator has to change the setpoint continuously. As already described in the previous section, the computer can also perform a regulating function. Indeed after comparing sensor-outputs to rated value points inside the memory of the machine, an on/off, a proportional, integral or differential type action or a combination can be calculated. The signals which represent these actions could actuate an appropriate control device such as a valve or a motor. This procedure is called Direct Digital Control (DDC). In contrast to DDC, Set Point Control (SPC) retains the conventional regulation chains and makes the computer operate on the setpoints of each regulator. Thus, the computer acts as a superregulator which sets the control profiles and is able to compensate for malfunctioning of the conventional controllers.

Already in 1973, Nyiri evaluated the use of DDC in fermentation processes. He stated that DDC could be found effective and economical in batch fermentations when a relatively large number of fermentors and a small number of control variables have to be controlled. Later on, as well setpoint control as direct digital control systems have been designed. Humphrey (1977) however stated that DDC is particular useful to plant scale fermentation systems and SPC to research and pilot plant scale fermentors, because SPC is modular and the instrumentation can be readily substituted as improvements are made. However in contrast with Humphrey (1977), Hampel et al.(1979) and Rolf et al. (1982a) state the additional flexibility of DDC, favouring DDC for pilot plants and research. Indeed, different controller parameters can be adjusted easily and the control algorithm itself can even be changed without new hardware. One of the great disadvantages of DDC is unfortunately the safety in case of a computer failure. Also, DDC sets a heavy load on the computer-system. With SPC, the computer is freed for more "intelligent" tasks, such as data-reduction, calculation of process-parameters and so on. For this reason, some groups made the option of SPC. Also in our laboratory SPC was chosen. In the last three years however one has noticed the upcome of the microprocessor. As these digital component is less and less expensive and as its performance is growing during the years one witnesses a trend towards "hierarchical control". This means that a supervisory computer is applied for data retrieval, data reduction, data storage, simulation and calculation of process status variables and control profiles and that this computer is connected to smaller microprocessor-systems for the ultimate "low-level" control of the control variables. This diminishes the load on the bigger computer for DDC as DDC is now performed by the smaller microprocessor-system. More and more authors are taking this option (Meiners et al., 1983 ; Rolf et al., 1982a, 1982b ; Bernard et al., 1983). Also in our laboratory, we try to implement the microprocessor in the lowest control function. Especially, because as BERNARD et al. (1983) also state that the direct regulation of control parameters gives a direct insight into culture physiology. This means that for instance the amount of acid - or base addition in pH-control is proportional to the metabolic activity of the micro-organism. Hampel et al. (1981) have even correlated this amount with the bacterial biomass for a fermentation with Arthrobacter species. So, the microcomputer here is both used as a control device and as a sensor ! More on this topic will be found in a later section.

A very recent development in computerized automation is the concept of distributed control. This concept is especially useful on a plant scale. It is realized nowadays that many control actions and managerial decisions are distributed over the entire plant. A new automation structure nowadays is seen as a highly sophisticated data link that connects many different small processors spread throughout the plant, each one in charge of its local control function. In that manner, the "bureaucratic" nature of stringent hierarchical organizations is avoided and instead a bus-, star- or network structure is provided which is able to provide the automated control functions in a much more natural way. A detailed design in the context of fermentation equipment remains to be done.

SPECIAL PROBLEMS

Though the hardware for computer-fermentor coupling is readily available now, there remain major problems in the field of computer automation of fermentation processes. One of the first to mention is the lack of sensors available nowadays to monitor the fermentation. Even very basic variables like biomass and product concentration are often difficult to monitor on-line. New developments will be discussed later. In addition little is known concerning the dynamics of the process. It has been more or less a habit in fermentation practice to grow organisms under steady-state or balanced-growth conditions, which are not particularly suited to discover the basic features of the cell behaviour and its potentialities. Consequently final optimization is still in its infancy.

The sensor problem

One of the greatest bottlenecks in computer-controlled fermentation is the lack of adequate sensors. Table 1 gives a overview of available sensors on the market today. As can be noticed, relatively few fermentation-related sensors are available - Higham (1983) emphasizes that nowadays the measurements of temperature, pressure, flow and level account for more than 90% of the measurements made in the process industries.

Table 1. Sensors, used in process-industry today

type of variable	measurement	instrument
physical	fermentor-temperature ambient temperature	thermistor
	vessel pressure barometric pressure	diafragma- pressure meter
	motor-energy consumption(Power)	-dynamometer -chain-meter
	agitation (revolution speed)	tachymeter
	mass flow rate (air)	-rotameter -electromagnetic airflowmeter
	flows of additives (feed, antifoam, acid or base, ...)	-electromagnetic flowmeter -weight
	antifoam detection	-conductivity-probe -capacitive-probe
	viscosity	viscosimeter
chemical	pH	combined glass- reference elec- trode
	dissolved oxygen	-amperometric probe -polarographic probe
	$\% \, O_2$ in inlet- and outletgasstream	-paromagnetic sensor -mass-spectrometer
	$\% \, CO_2$ in inlet- and outletgasstream	-infrared analyzer -thermal conducti- vity -mass-spectrometer

Table 2. Recent developments

type of variable	measurement	instrument
chemical	dissolved carbondioxide	gas-sensitive electrode
	redox	combined Pt-reference electrode system
	dissolved gases and volatiles (ethanol, methanol, ...)	mass spectrometer
	NH_4^+, Ca^{2+}, Cl^-,...	ion-specific electrodes
	glucose ureum lactose galactose ethanol cefalosporine penicillin G sucrose	-enzym-pH-electrode -enzym-O_2-electrode -enzym thermistor -enzym-transistor (ENFET) -ellipsometry -piezo-electrical crystals -opto-electronic sensor (color) -auto-analyzers (colorimetry)
biological	biomass	optical density meter (turbidity)
	NADH	fluorometer
	ATP, ADP, AMP	bioluminiscence-meter
	heat	calorimeter

Mass, density and measurements like conductivity, pH, ... account only for most of the remainder. So, upto now almost no sensors for the important physiological process variables as biomass, substrate concentration and productconcentration are on the market. Therefore much effort is made to fill up this gap. Two approaches can be applied. In the first place, much can be done by software. Indeed, one has implemented the "net effect" sensor. This is an algorithm that conditions and combines on-line measurements of physical and chemical variables to yield a derived quantity that is highly correlated with a biochemical or biological variable. The concept has been suggested by Nyiri (1972), Cooney et al. (1975) and

Spriet and Vansteenkiste (1978). More on this type of sensor will be
discussed in the second subsection. In the second place, a lot of
effort is put nowadays in the development of new sensors in hardware
terms. The next subsection gives an overview.

"Hardware" sensors

A whole range of different types of sensors have been built.
Table 2 represents some recent developments. Some of these examples
will be discussed in the following paragraphs.

Optical density

From table 2 one notices the very few detectors available for
biomass-measurement. For instance, optical density (turbidity)has
been studied as an on-line sensor for biomass in many cases. Roughly,
a relationship similar to the Beer-Lambert law of absorption exists
between the optical density and the biomass concentration at low
cell densities. As this relationship is only linear for OD readings
in the range 0.1-0.4, dilution is necessary (Lee et al., 1980).
Different ways exist to perform this dilution. Blachère and Jamart
(1969) builded a "biophotometer" which had a sensitivity of 0-10g/1
dry cell weight. In this device the photocell was located very close
to the photometric chamber, resulting in a short path-length and thus
in a dilution. Lee and Lim (1979) at the contrary constructed an
effective dilution device wherein the dilution was obtained without
interfering with the culture stream by insertion of a tube filled
with distilled water in the flow-through cell. By varying the inner
diameter of this tube they could perform different dilutions.
However looking into more detail into their data, one notices in
spite of the better linear relationship a lower sensitivity as a
small change in OD corresponds now to a large difference in dry cell
weight. As the previous devices demand a withdrawal of a sample,
some authors describe a submersible optical density-meter. Ohashi
et al. (1979) for instance made such device with a linear relationship (OD-DCW) up to 1.0 in optical density. Williams and Brain (1976)
made use of a different property of cell suspensions. They relied
on the back scatter of light on a photo-transistor, placed in a
bundle of flexible optic fibre. This device seems promising. A total
different approach can be found in the article of Wei et al. (1983),
who made use of separate small scale fermentors, run in parallel
with the full-scale fermentation, to determine the biomass. This was
necessary because they studied fermentations of yeast cells in semi-
solid gels, which hampered the sampling. Such technique may however
be doubtful for ordinary submersed fermentations because of scale-up
variability between different vessels. Fazel-Madjlessi and Bailey
(1979) studied the application of a laser flow microfluorometer as
another approach. This instrument permits rapid experimental determination of the compositions of individual members of a cell population. As the authors state, this technique can be valuable in identifying subpopulations of which the dynamics may be related to for
instance secondary metabolite production. This is particularly useful

for building structured models (see further). Finally, our laboratory has also tested the relationship optical density - dry cell
weight for various fermentations. To have some idea of the obtained
results we shall only describe two examples from a fermentation with
Bacillus brevis ATCC 9999 for the production of the antibiotic
gramicidin S. Both fermentations shall be used further in other sections. The following conditions and methods were applied. Both
examples were carried out in respectively 15 l and 18 l complex
medium, containing 3% yeast-extract and 3% bacto-peptone (pH adjusted
to 7.3), in a computer-coupled fermentor of 28 l working volume.
Both fermentations were conducted in batch mode and at free pH.
Example one was performed at a "high" oxygen level (% DOT > 10%) by
controlling the air-flow rate and the second at a "low" oxygen level
(% DOT dropped to $\pm$ 0%). In the second example the air flow rate
was kept constant. As the gramicidin S production depends on the
oxygen level, the first example showed no production and the second
one had a maximum titer of 1200 mg gramiciden S/l. More details of
these fermentations can be found in Vandamme et al. (1981, 1982).
Figures 2 and 3 represent the correlation OD-DCW (g/l) respectively
for example one and example two. The optical density was obtained
after careful dilution in order to obtain a reading between 0.1 and
0.3. In Fig. 2, one notices only a linear relationship for the data
in the exponential growth-phase. Moreover, the data-pairs of the
transition-phase are situated "above" this linear regression in
contrast with the fact that normally higher cell-densities result
in a lower extinction due to secondary scattering. Possible explanations may be : (1) As growth continues, foaming increases, if chemical antifoam compounds are added, the number of particles in the
medium increases due to the formation of small spheres of antifoam ;
(2) The total scattered light is greater, the greater the difference
between the refractive index of the medium and of the particles
(Pirt, 1975). As the fermentation proceeds, some metabolites (e.g.
NH_3, amino acids, short peptides, ...) are released, resulting in a
change in the refraction index.
In Fig. 3 one notices a linear relationship for all phases. Only the
dilution 1/75 showed an appreciable deviation, probably due to a
systematic error. Another very surprising factor is the difference
in the slope of the relationship between example 1 and 2. A statistical test on the identity of both regression-curves showed that
they were not identical at all. So, in this case an OD-measurement
is not very useful as an "absolute" estimator of the biomass.
Though OD-measurements have been with us for a long time, there remains a lot to be done in order to properly evaluate its potential.
A microprocessor here might be of help in providing means for tuning
the relationship and linearize non-linear trends.

Fluorometry

Another approach is the use of a fluorometric measurement as
an estimator of the biomass. Beyeler et al. (1981) for instance report a sterilizable fluorometer which could monitor growth.

200

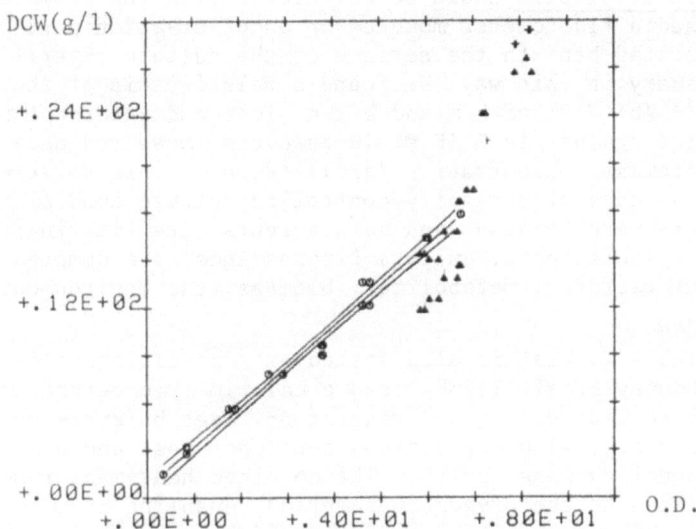

Figure 2 : Correlation between DCW (g/l) and O.D. for example one :
O = exponential phase ; + = transition phase ;
Δ = stationary (lysis) phase.

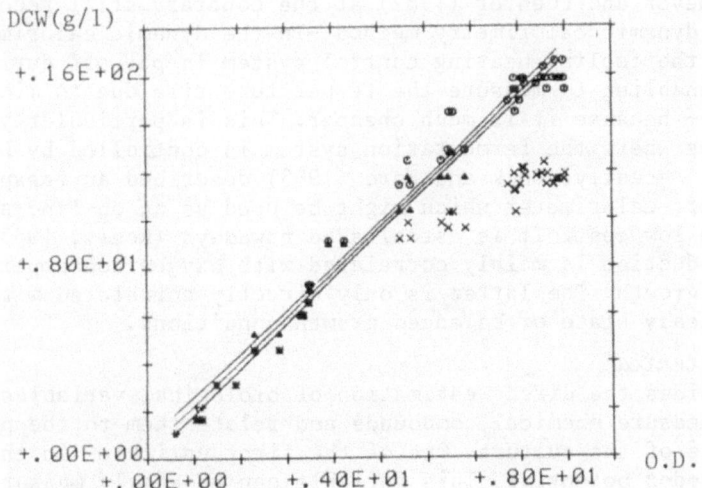

Figure 3 : Correlation between DCW (g/l) and O.D. for example two :
+ = 1/6 ; ✕ = 1/15 ; Δ = 1/30 ; O = 1/60 ; X = 1,75 :
dilutionfactor.

Indeed, the NAD(P)H-dependent fluorescence signal during batch growth of __Candida tropicalis__ could be correlated with the biomass. Zabriskie (1979) used a fluorometer mounted on an observation port of the fermentor located beneath the surface of the culture. Sterilisation was not necessary in this way. He found a relationship of the form :
$X = \exp(-m/b) \; F^{1/m}$ with m and b the slope and intercept when ln F was plotted against ln X (F = fluorescence (measured as a potential) and X = biomass concentration (g/l)). However this relationship holds only in the case of carefully controlled culture conditions (pH and temperature were constant and no nutrients were limiting).
In other circumstances, culture fluorescence is a complex function of several factors : metabolites, biomass, the environment, ...

Calorimetry

In Table 2, heat is also stated as a possible biological sensor. Indeed, Cooney et al. (1968) used a calorimetric method to determine the produced heat during a fermentation. Heat balances over the fermentation system with corrections for vapor loss and heat loss to the surroundings were applied. The obtained heat-measurement could be correlated to the oxygen consumption, enabling them to use oxygen consumption for a possible prediction of heat evolution. Mou and Cooney (1976) used approximately the same method as Cooney et al. (1968) but they claimed that thermal data may prove significant in monitoring biomass during the growth phase. Also Marison and von Stockar (1983) found a remarkably close correlation between heat generation and cell density. Goma and Ribot (1975) have even taken a patent on a procedure for measuring heat as a means to monitor biomass. Moreover, nowadays some special devices for measuring microbial heat during the fermentation exist on the market (Eriksson, 1981). Beyer and Fuehrer (1982) at the contrary still recommend to use the dynamic calorimetry method -In the dynamic calorimetric method, the cooling/heating control system is put off during a short period enabling to measure the temperature rise due to microbial activity- because it is much cheaper. This is particularly true in the cases where the fermentation system is controlled by DDC-means! However, recently, Lock and Ford (1983) described an inexpensive flow micro-calorimeter which might be used as an on-line sensor at relative low cost. It is established nowadays (Roels, 1983) that heat production is mainly correlated with oxygen consumption in aerobic growth. The latter is only directly correlated with growth under steady state or balanced growth conditions.

Redox potential

Besides the direct estimation of biological variables, one can try to measure chemical compounds and relate them to the physiological state of the culture. One of the first variables in this context is the redox potential. This variable can be easily measured by a platinum electrode and a reference electrode. Kjaergaard and Joergensen (1979) claimed that redox-potentials yield more information about the oxidative status in aerobic or partially aerobic microbial cultures than a dissolved oxygen measurement. Indeed, according to Dahod (1982) redox probes have a higher range due to the availability of negative redox potentials where DO is zero. A disadvantage however is the dependence of the redox potential on pH. Fortunately, pH-monitoring is so widely spread that the redox-signal can easily be corrected instanteneously for broth-pH. Finally, Heinzle (1981b) even describes the use of such a redox electrode in analyzing dissolved hydrogen.

Mass spectrometry

Another very promising instrument is the mass spectrometer for analyzing gas partial pressures, dissolved gases and volatiles. Especially, because present quadrupole instruments are small, easy to handle and relative low in price. Linear responses are found over a wide range of concentrations. The instrument is very power-ful if it is interfaced to a computer system, as described by Heinzle (1981a) and Heinzle et al. (1981b), Pungor et al. (1980), Fish et al. (1981). In all cases, the tube method is used. In this method, the volatiles diffuse through a teflon or another kind of membrane into a carrier gasstream, which brings the compound, to be measured, to the analyzing instrument. As stated by Heinzle (1981a), response characteristics of the membrane probe are very much dependent on the membrane and on the compound analyzed.

Auto-analyzers

As it is not always possible to monitor certain status variables within the fermentor, automatic analyzers are developed to perform the desired analysis outside the culture. These instruments can be based on colorimetry or turbidimetry. The sample is conditioned by adding buffers and reagents to the sample before it is entered in the measuring chamber. Examples are automatic monitoring of protease activity (Leisola et al., 1979a), cellulose activity (Leisola and Virkkunen, 1979b) and α-amylase and gluco-amylase activity (Leisola et al., 1980). Interesting here is also the control of the automatic analyzer system by a small microcomputer. Indeed, the microprocessor can easily control the sampling of the culture, the addition of reagents, recalibration of the instrument, and calculation of the ultimate concentration of the compound.

Ion-selective electrodes

As can be noticed in Table 2, ion-selective electrodes (among which also gas-sensing electrodes and enzyme-electrodes) are very useful to monitor CO_2, NH_3, some anorganic ions and most important some product- and substrate concentrations (glucose, penicilline, cephalosporine, sucrose, ...). An ion-selective electrode consists of a membrane, which is selective permeable for a restricted number of species, in the ideal case only for the species wanted. Over this membrane a potential is exhibited dependent on the different activi-ties of the specific ion on both sides of the membrane. The measure-ment of this electrical potential difference demands a galvanic loop. Therefore in the internal solution of the ion-selective elec-trode, a reference electrode (usual a Ag/AgCl electrode) is brought and connected to an external reference electrode (cal/omel or Ag/AgCl electrode) in the sample-liquid. The potential across the selective membrane is given by the famous Nernst equation. In this equation, one notices that the potential will change linearly with the logarithm of the activity of the selected species. This offers a constant relative accuracy over a wide measuring range, which is a very favourable property. However, from this equation it is also clear that the relationship holds for "ion activities" and not directly for concentration units ! Therefore, in most cases, it is necessary to work at a constant ionic strength in order to measure concentrations. Another disadvantage of ion-selective electrodes, which also occurs in other analytical methods, is the interference with other constituents of the sample. The interfering species have to be removed prior to the measurement. At the contrary, the response is not affected by color, turbidity, volume, ... and the ion-selective electrode is sensitive, mechanically robust and non-destructive. An article on the impact of ion-selective electrodes on biotechnology has been published by Kell (1980). In the following paragraphs we shall discuss some ion-selective electrodes into more detail.

Hill and Thommel (1982) and Le Duy and Samson (1982) for instance describe the use of an ammonia electrode for ammonia nitrogen measurement, respectively in the methanogenic sludge and in a baker's yeast fermentation. As this electrode is of the gas-sensing type, it is necessary to raise the pH of the sample to more than 11.5 in order to expel the ammonium ions as ammonia. Therefore, the determination has to be performed on a separate sample if ammonium ions are to be determined. Thus, in this case, it is not possible to bring the ammonia probe into the culture liquid in the fermentor vessel. The same disadvantage occurs if one wants to determine bicarbonate and carbonate with a carbondioxide electrode. However if one wants only to quantify ammonia or carbondioxide gas, one can submerge the electrode directly into the culture liquid. In this respect, up to now, only one firm presents a sterilizable carbondioxide electrode which seems promising. A total different approach can be found in a publication of Karube and Suzuki (1983). They claimed that it is better to use an electrochemical determination of ammonia instead of a potentiometric detection as with an ammonia probe. Therefore they use immobilised cells of Nitrosomonas and Nitrobacter species, which convert the ammonia with the uptake of oxygen to respectively nitrite and nitrate. The amount of oxygen uptake, determined with an oxygen electrode, is a measure for the ammonia concentration. In contrast with the potentiometric ammonia electrode volatile compounds, such as amines, do no interfere with the determination.

A very similar approach as Karube and Suzuki (1983) is the development of a "microbial" sensor for nitrilotriacetic acid (NTA). Washed cells of Pseudomonas sp. were held close to the surface of the gas permeable membrane of an ammonia electrode by means of a dialysis membrane. The measurements were made in a thermostated cell and the nitrilotriacetic acid in the sample was utilized by Pseudomonas sp. as sole source of carbon and nitrogen with the formation of ammonia. In this way Kobos and Pyon (1980) demonstrated the ability of whole microbial cells to mediate reactions at a membrane electrode for analytical determinations.

Besides microbial cells, immobilized enzymes have been and are still extensively studied as mediators. Excellent reviews on this material have been published by Guilbault (1980) and Solsky (1983). It is generally accepted that an enzyme in the immobilized form can be used repeatedly over a reasonable period of time. Moreover the resulting "enzyme electrode probe" is in most cases very sensitive. However, some problems with these devices are (1) the sample must be conditioned ; it is not yet possible to bring a enzyme electrode directly into the fermentation broth ; (2) K_M (Michaelis-Menten constant) is frequently too low causing these electrodes to saturate at high concentrations of the substrate. According to Turner (1983) this last disadvantage can be remedied by producing diffusion limited probes with high enzyme loading factors. Normally, this results in an increased response time. Turner (1983) however, describes an electrode, which was constructed with carbodiimide immobilized glucose oxidase in the presence of ferrocene. The upperlimit of the linear response increased from 15 mM to 40 mM in spite of the fact that the response time remained below 30 seconds. Besides Turner (1983), several others have tried to improve the glucose - enzyme electrode, which Updike and Hicks introduced already in 1967. Enfors (1981a,b) for instance describes a glucose sensor based on a galvanic oxygen electrode with an electrolysis anode. The enzymatic consumption of oxygen by the glucose oxidase is utilized to control electrolytic generation of oxygen from water so that this enzyme-electrode operates under ambient oxygen tension or even in oxygen free broths. In this manner, the measuring range of this electrode is extended since it will no longer suffer from oxygen limitation. Enfors (1981a,b) describes the possibility of this electrode to submerge it in the fermentation broth. Finally, Kernevez et al. (1983) describe a computerized glucose enzyme electrode, settled in a cell outside the vessel. The whole measuring cycle as air rinsing, introduction of the sample, water rinsing and calibration procedure is now set under software control. Besides glucose electrodes, numerous other enzyme electrodes are described, such as penicillin-sensitive electrodes (Kell, 1980), amino-acid enzyme electrodes (e.g. L-lysine electrode (Tran et al., 1983), alcohol electrodes (Verduyn et al., 1983), and so on. Recently, Wingard (1983) points to the fact that some oxidation-reduction enzymes are highly complex spatial arrangements of sequences of several cofactors and associated proteins. For instance, Liu and Chen (1982) studied gel-immobilized oxidoreductases with NADH as cofactor and used ferricyanide ions as electron mediators to couple the oxidation to the reduction reaction. Finally, it is worthwhile to point to some other developments : (1) Wingard (1983) describes an electrode wherein the electron conducting membrane becomes a continuum with the cofactor instead of with the enzyme. This requires an immobilization of the cofactor in such way that the apo-enzyme can form a complex with the cofactor and the substrate to perform the reaction. Wingard (1983) has coupled in this respect riboflavin to glassy carbon, which seems a promising development ; (2) besides enzyme electrodes, one has also studied "substrate probes" with the substrate immobilized close to the electrode for monitoring the enzymes themselves. One of the problems encountered with this type of sensor is that the substrate is used up. Moreover the enzyme reaction must be monitored dynamically and not in steady state conditions. Guilbault (1980) has given a short overview of this type of sensor.

Besides ion-selective electrodes, integrated ion-sensitive sensors on small silicon chips, thermistors (heat) and some special techniques have been used for measuring the enzyme reaction. These will be discussed in the next paragraphs.

Enzyme thermistor

Similar to the use of heat generated by growth as a means to estimate biomass, heat, generated by enzymatic reactions, can be applied for measuring various substances. Therefore, one has developed the enzyme thermistor. This instrument consists of an automatic sampling circuit, to take and condition the sample and a calorimeter where the heat of the enzyme reaction is monitored. Several substances, among which the enzymes themselves, can be monitored in this way (Mosbach et al., 1983 ; Danielsson et al., 1981 ; Mandenius et al., 1981). A very successful application was the determination of penicillin in fermentation broths, analyzed with a unit containing penicillinase (β-lactamase) immobilized on controlled porous glass beads. Several units have been installed in the antibiotics industry (Mosbach et al., 1983).

Metal oxide semiconductor sensors

In the last four to five years, much progress has been made in integrated circuit fabrication. One of the advanced techniques is planar processing. This procedure has been proved successful in the development of several semiconductor sensors (Sansen, 1983). A whole range of different types have been made : pressure sensors, optical sensors, ion-selective field effect transistors (ISFET), enzyme transistors (ENFET) and so on. Kempe and Schallenberger (1983) and Vorlop et al. (1983) describe an alcohol sensor based on semiconductor principles. When combustible or reducing gases or organic solvent vapor adsorb on the surface of sintered $Sn\ O_2$ then the electrical resistance markedly decreases. The sensor is very stable and had an average life time of several years. Besides such a gassensor, undoubtly ISFET's and ENFET's may prove to be the devices of the future. The coating on top of the gate of a FET absorps ions out of the solution, resulting in a current through the transistor. The theory has been described by Bergveld and De Rooij (1979) and Croset (1982). Similar as enzyme electrodes, ENFET's make use of immobilized enzymes and the modified FET is applied to measure the enzyme activity. Danielsson et al. (1983) and Winquist et al. (1981) describe the use of MOS (metal oxide semiconductor) structures with a thin coating of palladium, resulting in a sensor sensitive to gaseous hydrogen. Hydrogen dehydrogenase was immobilized with this structure enabling them to measure NAD^+/NADH. In this way all reactions involving these cofactors can be determined, as for instance the determination of pyruvate (Danielsson et al., 1983).

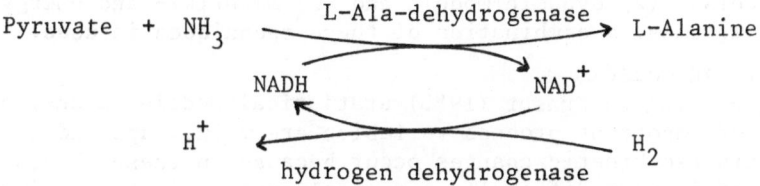

Moreover, Danielsson et al. (1983) deposited small amounts of iridium on the palladium film in order to measure ammonia directly. However, as these MOS-structures are usually kept at high temperatures (150°C) to prevent the sticking of water molecules to the metal surface and to speed up the response time, they can not yet be submerged into the broth although Danielsson et al. (1983) are investigating this possibility now.

Some special techniques

Mosbach et al. (1983) describe some other techniques, ellipsometry and the use of piezo-electric crystals, which however have not yet proven to be valuable in monitoring fermentation processes. Lowe et al. (1983) developed a solid-phase opto-electronic device, comprising a biospecific ligand covalently attached to a transparant cellophane membrane, brought between a photo-diode and a light emitting diode in a flow through cell. For instance penicillinase from Bacillus cereus was co-immobilized with bromocresol green. On contact with penicillin protons were generated by hydrolysis of the β-lactam ring, causing the bromocresol green to change from blue-green to yellow and thus a raise in output-voltage from the photodiode. Also this technique seems promising although the same problems as with auto-analyzers occur.

"Net-effect" sensors

As long as the previous described "hardware" sensors are not available on the market and reliable to use in an industrial environment, one has to turn for the determinations of biomass substrate concentration and productconcentration to the "net-effect" sensors. Indeed, it is possible with a digital machine to estimate these important physiological variables merely from the variables of Table 1. Humphrey (1977) can be quoted in this respect : "It is my feeling that the most significant role for computers in fermentation systems will be their use in biomass and substrate monitoring by indirect means through component balancing. The reasons for this belief stem from the fact that adequate on-line sterilizable biomass and substrate sensors have not been developed in spite of the fact that since the mid-1940's the fermentation industry has indicated that this is one of the major limitations to fermentation process control". In fact, despite the developments mentioned in the previous subsection, net-effect sensors, sometimes also called "derived quantities" or "gateway" sensors or "synthetic quantities", are still extensively used and investigated. It has now been established that their construction is based on three techniques : (1) statistical models ; (2) dynamic models and (3) material- and energy balances. Sometimes, a combination of these techniques is used.

Statistical models

According to Fraser (1983) statistical models as on-line estimators of important process variables are mainly applied in cases where drastic kinetic changes occur because in these phases dynamic modelling is very difficult. A typical example is the estimation of biomass from the amount of acid or/and alkali added for pH-control. As mentioned already in a previous section, Hampel et al. (1981) found a linear regression relationship between the amount of acid or base added and the biomass concentration. Enfors (1981) and Dostálek (1975) describe even an instrument, "the dose monitor", which is commercially available now and which could continuously monitor the dosing by pumps. They also mentioned that the relationship only holds in the cases where : (1) the sole nitrogen source is ammonia ; (2) energy metabolism yields a constant production of acids ; (3) amino acids are partly consumed either as carbon/energy source or as ammonia source, leaving either ammonia or an acid and most important (4) the yield coefficient is constant, which is particularly true in the exponential growth phase. In the next paragraphs those constraints are sometimes also applied. Also Heijnen (1981) points to a certain caution when estimating biomass from acid- or base addition.

Dynamical models

Recently, an excellent book on the methodology of the construction of models in biotechnology has been published by Roels (1983). In combination with the book of Spriet and Vansteenkiste (1982), one has the tools to simulate the biological phenomena on a computer In this paragraph, we shall discuss the use of these models as on-line "net-effect" sensors. According to Roels (1983) one starts with a balance equation. For an extensive quantity the relationship "accumulation = transformation + transport" holds. Applied to chemical reactions and with concentration as intensive quantity, one gets

$$(\underline{c}.V) = \underline{r}.a.V + \underline{f}.V \tag{1}$$

with $\underline{c} = [c_1 \ldots c_i \ldots c_n]$: concentrations of the n chemical components

$\underline{r} = [r_1 \ldots r_j \ldots r_m]$: reaction rates of the m independent reactions

$a = [a_{ij}]$: stochiometry

$$\underline{f} = [\, f_1 \, \ldots \, f_i \, \ldots \, f_n \,] \quad : \quad \text{fluxes of the n chemical compounds}$$

In batch mode there is no transport from and to the environment, ex-cept for the inlet and outlet gasstream. In this case $\underline{f} = \underline{0}$. So, if $\underline{r}$ and a are known, one can determine the kinetic equations. In con-tinuous culture, one obtains at steady state conditions and constant volume : $(\underline{c}.V) = \underline{0}$, which means $\underline{f} = -\underline{r}.a$. So, if the rates $\underline{r}$ and a are known, one can estimate immediately the fluxes $\underline{f}$. Thus, in both cases one has to determine the stochiometry matrix and most important the rates $\underline{r}$. In many cases these rate equations are empirical. Roels and Kossen (1978) give a review of models used in microbiology and biochemistry. Table 3 gives an overview of some well known "growth" equations. All these equations are descriptions of unstructured models which means that no attempt has been made to distinguish different components within the biomass. In other cases, one has tried to split up the biomass in several subunits and constructed structured models. Roels (1983) gives a methodology to build these models. Esener et al. (1982) point to the difficulties, encountered in structured modelling, namely e.g. determination of the internal composition is a very difficult task and very prone to errors. However, as already pointed out, microfluorometry can be a possible solution, especial-ly if it could be applied on-line. Further research is necessary. Until more progress is made, one has to rely on unstructured models. Zabriskie et al. (1976,78) describe the use of an oxygen balance and the linear substrate consumption law as a means to estimate growth. Their equation was :

$$\beta.\text{OUR} = m_{O_2/X} \cdot x + \frac{1}{Y_{X/O_2}} \cdot \frac{dx}{dt} \qquad (2)$$

with β = metabolic function, equals 1 when the glycolysis and the TCA-cycle is followed in the metabolism

OUR = oxygen uptake rate, calculated $(\frac{\text{gr } O_2}{1 \text{ min}})$ from a balance

$m_{O_2/X}$ = maintenance coefficient (gr O_2/gr biomass.min)

Y_{X/O_2} = growth-yield coefficient on (gr biomass/gr oxygen)

Also Humphrey (1977), Alford (1978), Hampel (1980), Hampel et al. (1979) and Holmberg (1981) and Holmberg et al. (1980) have described a similar relationship. After the determination of m and Y one could on the basis of the consumption of a substrate (e.g. O_2-uptake) or the accumulation of a metabolite (e.g. CO_2-release) estimate the biomass concentration, growth rate and substrate consumption. Much care however has to be taken in applying these kinds of models be-cause the assumptions which underly the equation will not in all cases serve. Moreover, reinitialization during the course of a fer-mentation may be necessary.

Material- and energy balances

One of the most developed technique for the construction of "net-effect" sensors is certainly the use of mass- and energy balances.

Table 3. A survey of some unstructured models

Author	Growth equation	Remark
Monod	$\dfrac{dx}{dt} = \mu_m \cdot \dfrac{S}{K_S + S} \cdot x$	growth limited by one substrate
Konak	$\dfrac{\partial \mu}{\partial S} = k(\mu_m - \mu)^P$	growth rate dependent on the substrate concentration
Blackman	$\mu = \dfrac{S}{a} \quad$ for $S < \mu_m \cdot a$ $\mu = \mu_m \quad$ for $S > \mu_m \cdot a$	growth rate dependent on the substrate concentration
Powell	$\mu = \dfrac{\mu_m(K_S + L + S)}{2L} \left(1 - \sqrt{1 - \dfrac{4 \cdot L \cdot S}{(K_S + L + S)^2}}\right)$	mass transfer to and within the organism is combined with Monod kinetics
Moser	$\mu = \mu_m \cdot \dfrac{S^\lambda}{K_S + S^\lambda}$	growth rate dependent on the substrate concentration
logistic law	$\dfrac{dx}{dt} = \mu_m \cdot x \cdot \left\{1 - \dfrac{x}{x_m}\right\}$	no direct relationship between growth and the concentration of the substrate is assumed

x	= biomass concentration
μ	= specific growth rate
μ_m	= maximum specific growth rate
S	= substrate concentration
K_S	= the Monod-constant (or Saturation constant)
k, p, a, λ	= constants
L	= $\dfrac{q_{max}}{F}$
q_{max}	= maximum of the specific substrate consumption rate
F	= surface area for mass transfer

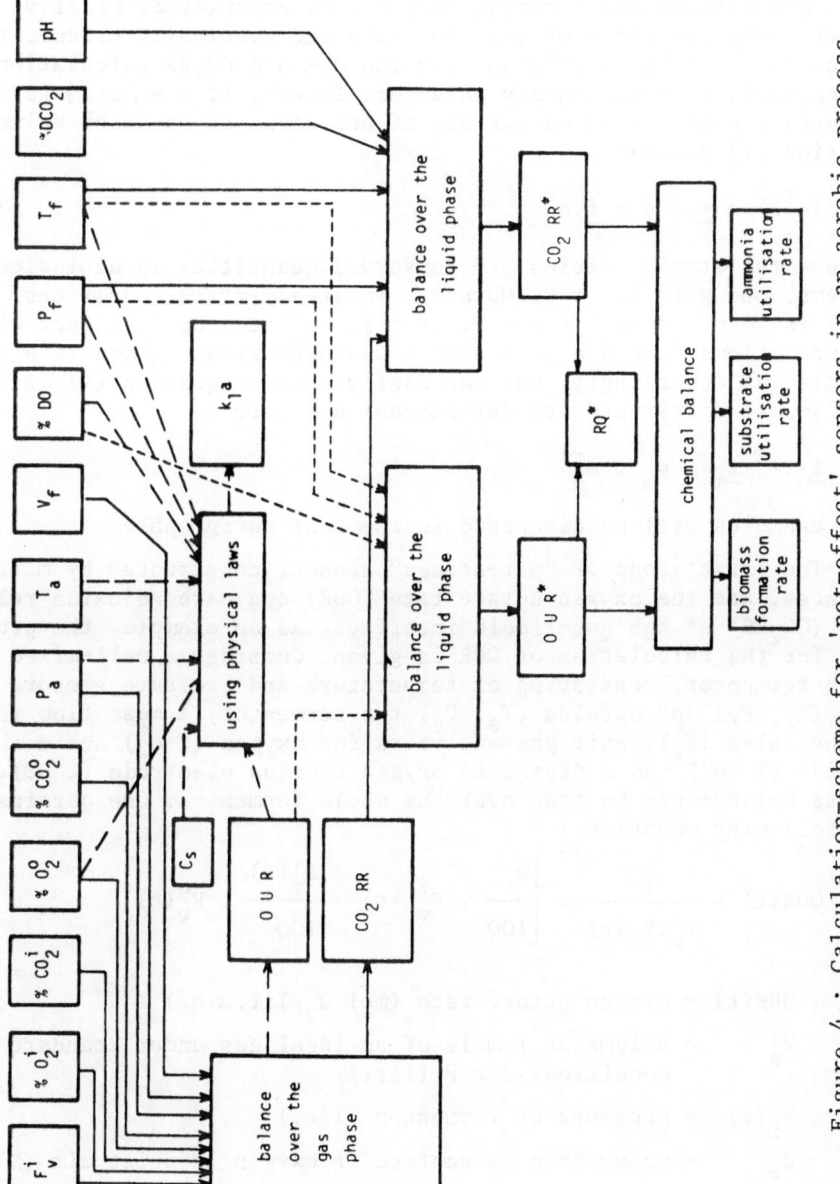

Figure 4 : Calculation-scheme for 'net-effect' sensors in an aerobic process

Nagai (1979) gives an excellent summary in this respect. Fig. 4 gives
an overview of the calculation-scheme, applied to aerobic fermentations. Via ordinary mass balances, OUR (oxygen uptake rate), CO_2RR
(carbondioxide release rate), volumetric oxygen transfer coefficient
(k_1a) and respiratory quotient (RQ) can be calculated. Finally,
growth rate, substrate utilization rate and eventually product formation rate can be calculated from the OUR and CO_2RR calculations
by application of elementary balances. Indeed, if $e = [e_{ij}]$ is the
elementary composition matrix and if one works at constant volume,
equation (1) becomes :

$$(\underline{\dot{c}}.e) = \underline{r}.a.e + \underline{f}.e \qquad (3)$$

Because the atomic species are conserved quantities in biological
systems, one gets a.e = 0. Moreover in steady-state conditions,
$(\underline{\dot{c}}.e)$ is zero, so that $\underline{f}.e = \underline{0}$. If $\underline{f}$ is splitted up in a part with
measured flows $\underline{f}_m$ and a part with unmeasured flows $\underline{f}_u$ and if e is
splitted up accordingly, one can easily derive equation (4) if e_u
is a square matrix and its determinant not zero :

$$\underline{f}_u = -\underline{f}_m \cdot e_m \cdot e_u^{-1} \qquad (4)$$

Some examples will be discussed in the next paragraphs.

The first group of "net-effect" sensor, constructed by mass-
balances, are the oxygen uptake rate (OUR) and carbondioxide release
rate (CO_2RR) at the gas-liquid interface. As an example, the procedure for the calculation of OUR is given. Consider a well-mixed
batch fermentor, consisting of temperature and pressure sensors inside (T_f, P_f) and outside (T_a, P_a) the fermentor, a mass flow rate
at the inlet (F_v^i), exit gas analyzers for oxygen ($\% \, O_2^o$) and carbondioxide ($\% \, CO_2^o$) and a dissolved oxygen tension electrode ($\%$ DO).
A mass balance can be made over the whole fermentor. One obtains
the following equation :

$$OUR(t) = \frac{1}{V_s' \cdot V_f(t)} \cdot \left[\frac{a_v}{100} \cdot F_v^i(t) - \frac{a_v'(t)}{100} \cdot F_v^o(t) \right]$$

with : OUR(t) = oxygen uptake rate (mol O_2/lit.min.)

$\quad V_s' \qquad$ = volume of 1 mole of an ideal gas under standard
$\qquad\qquad$ conditions for F_v(lit.)

$\quad V_f(t) \quad$ = contents of fermentor (lit.)

$\quad a_v \qquad$ = volumetric percentage of oxygen in inlet air ($\%$)

$\quad a_v'(t) \quad$ = volumetric percentage of oxygen in outlet air ($\%$)

$\quad \% \, O_2^o \quad$ = volumetric percentage of oxygen in outlet as measur-
$\qquad\qquad$ ed by the oxygen analyzer ($\%$)

$\%\ CO_2^o$ = volumetric percentage of carbondioxide in outlet as measured by the carbondioxide analyzer (%)

$F_v^i(t)$ = gas flow rate at standard conditions in the inlet (lit./min.)

$F_v^o(t)$ = gas flow rate at standard conditions in the outlet (lit./min.)

To obtain a better estimate of the outlet flow rate one can utilize the carbondioxide concentration in the outlet gas stream, measured by the CO_2-analyzer ($\%\ CO_2^o$). In addition the calibration conditions (T_a^c, P_a^c) of the analyzers may be taken into consideration. The oxygen analyzer is often of the paramagnetic type. It is temperature controlled but the reading is dependent on barometric pressure (P_a). The carbondioxide analyzer is usually an infra-red measuring device and has to be corrected for ambient temperature (T_a) and barometric pressure (P_a). Finally the gas flow rate is measured at the inlet while the gas analysis occurs after conditioning in the outlet causing a time lag (τ). If the inlet- and outlet gas has been dried, a more precise equation for the oxygen uptake rate can be obtained. The derivation has been reported elsewhere (Spriet, 1980). One finds :

$$OUR(t) = \frac{F_v^i(t)}{100.V_s'.V_f(t)} \cdot \frac{100.\,a_v.P_a^c - \%O_2^o(t').P_a(t')\,.T_a(t')+f(t')}{g(t')}$$

$$f(t') = b_v.\%O_2^o(t').P_a(t').T_a(t') - a_v.\%CO_2^o(t').P_a(t').T_a^c$$

$$g(t') = 100.P_a^c.T_a(t') - \%O_2^o(t').P_a(t').T_a(t') - \%CO_2^o(t').P_a(t').T_a^c$$

$$t' = t + \tau$$

with : P_a, T_a = ambient pressure (mm Hg) and temperature ($^\circ$K)

P_a^c, T_a^c = ambient pressure and temperature at calibration time

b_v = volumetric percentage of carbondioxide in inlet(%)

The carbondioxide release rate can be calculated in a similar way. Then balances can be made over the biomass-liquid interface to obtain OUR^* and CO_2RR^*. However, in contrast with OUR^*, where the necessary dissolved oxygen concentration can be measured easily, carbondioxide and most important bicarbonate and eventually carbonate-ions are very difficult to determine. Therefore, up to now, one could never estimate CO_2RR^*. It is hoped that this problem can be solved in the near future.

From the physiological variables OUR^* and CO_2RR^* (in most cases CO_2RR is used), one can calculate the respiratory quotient (RQ) which is :

$$RQ = \frac{CO_2RR^*}{OUR^*} \qquad \text{with } OUR^* \text{ and } CO_2RR^* \text{ in moles/lit.min.}$$

Nyiri et al. (1975) applied this variable as an indicator of the physiological condition of the cell. Indeed, Wang et al. (1977) established RQ as a quantitative indicator of ethanol formation in a baker's yeast fermentation. Moreover, they described the use of this "net-effect" sensor as a control variable. Other authors used this same variable as well to control the baker's yeast fermentation (Sumitani et al., 1983 ; Whaite et al., 1978 ; Ramirez et al., 1981).

From the variable OUR, one can also calculate the oxygen mass transfer coefficient k_1a, which governs the transfer of oxygen from the gas phase to the liquid phase. This variable indicates how well oxygen is supplied to the culture medium. Mainly two techniques are available for the estimation of k_1a. Spriet et al. (1982) described the theory and the differences between the dynamic measurement method (or "gassing-out" method) and the static method. Basically in the "gassing out" method a step in the aerating conditions is given and the dissolved oxygen electrode response monitored. From these data a k_1a-estimation can be performed. Indeed, the following formula can be applied (Nyeste et al., 1981) :

$$c(t) = - \frac{1}{K_1a} \cdot \frac{d\ c(t)}{dt} + c^*(t) + x.r_{O_2}$$

with : $c(t)$ = actual dissolved oxygen concentration (mg O_2/l)

K_1a = oxygen mass transfer coefficient (1/min)

$c^*(t)$ = oxygen concentration in the fermentation liquid under equilibrium (mg O_2/l)

x   = biomass concentration (g/l)

r_{O_2} = specific oxygen consumption of the micro-organism (mg O_2/g biomass).

It is evident that by this technique the process is disturbed. Therefore Spriet et al. (1982) proposed a non-interfering technique, particularly based on a mass balance over the fermentor. They obtained :

$$k_1a(t) = \frac{OUR(t)'}{\{c^*(t)-c^*(t_c)[\ \%DO(t)/100]\ \}}$$

with : $k_1a(t)$ = volumetric oxygen mass transfer coefficient (1/min)

$OUR(t)$ = oxygen uptake rate (mg O_2/l.min)

$c^*(t_c)$ = oxygen concentration in the fermentation liquid under equilibrium at calibration time (mg O_2/l)

$\%\ DO(t)$ = dissolved oxygen tension.

As outlined in the previous paragraphs, OUR(t) can be estimated from on-line measurements. Therefore with this method instantaneous k_1a estimations can be easily calculated by the on-line computer. Spriet et al. (1982) give some examples. One disadvantage of this approach however is the rather poor estimations in the beginning of the fermentation due to a large error in OUR. Indeed, given the finite precision of the sensors, the best results are obtained for large oxygen uptake rates. A very favourable property of the static method is the availability of an instantaneous estimation of k_1a during the fermentation, rather than delivering only one value in the case of the dynamic method. In addition an average value over the whole fermentor is found in that manner. In fact k_1a-estimations may prove feasible and useful in checking operator actions on flow rate and agitation speed. Moreover according to Jarai (1972) the value of k_1a seems to decrease continuously to the end of a secondary metabolite fermentation. At the contrary, in the case of extracellular enzyme fermentations the k_1a shows a tendency to increase. Although not yet very convincing, in fermentations at our laboratory such observations could also be found. For instance, in Fig.5 the k_1a estimates of the previously described fermentations of example 1 and 2, are represented. As example 1 showed no gramicidin S production, k_1a stayed very constant till the end of the fermentation, where as in example two k_1a decreased with antibiotic formation. A possible explanation may be the dependence of oxygen transfer on the rheological properties of the broth (Jarai, 1972). Further research in this direction is certainly necessary.

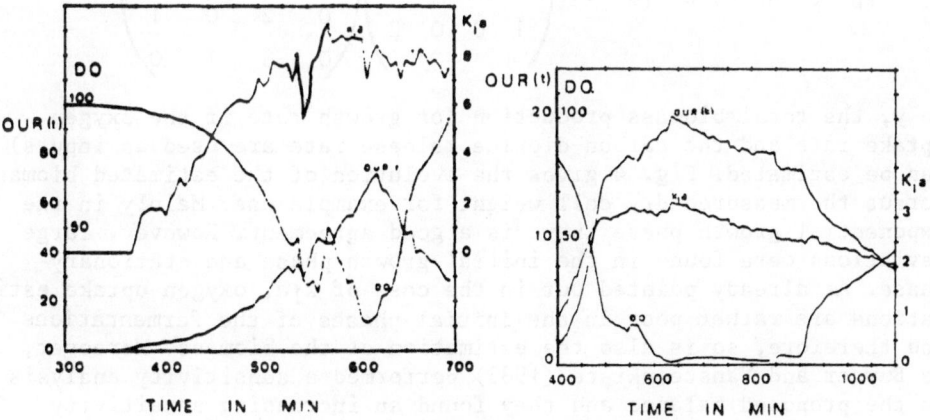

Figure 5 : $k_1a(t)$, OUR(t) – estimations and DO-measurements, respectively for example one and two. More details can be found in Spriet et al. (1982)

215

Finally, "net-effect" sensors can also be obtained from chemical considerations. One starts with a chemical representation of the growth process. For instance, if the substrate is utilized as the carbon and nitrogen source and if the process is aerobic of nature and if no products are formed, growth can be represented as:

$$p\ C_a\ H_b\ N_c\ O_d + x\ O_2 \rightarrow y\ C_\alpha\ H_\beta\ N_\gamma\ O_\xi + z\ CO_2 + u\ H_2O + v\ NH_3$$

substrate   biomass

with :   p = total substrate utilization per unit volume (mol/l)
        x = total oxygen uptake per unit volume (mol/l)
        y = total biomass production per unit volume (mol/l)
        z = total carbon-dioxide release per unit volume (mol/l)
        u = total water formation (or utilization) per unit volume
                                       (mol/l)
        v = total ammonia-release per unit volume (mol/l)
 a,b,c,d : elementary composition of the substrate
 α,β,γ,ξ : elementary composition of the biomass

By writing down the elemental balances, one has four mathematical constraints (C, H, N and O-balance). Moreover there are 6 variables (p, x, y, z, u, v) and 8 parameters (a, b, c, d, α, β, γ, ξ). If the parameters (elemental compositions) are known and if two of the six variables could be measured, one could estimate the other variables. Mathematically, this is expressed in equation (4). Indeed, if as in Fig. 3, the oxygen uptake x and carbon-dioxide release z are determined, equation (4) becomes :

$$(p\ \ y\ \ u\ \ v) = -(x\ \ z)\cdot \begin{pmatrix} 0 & 0 & 0 & 2 \\ 1 & 0 & 0 & 2 \end{pmatrix} \begin{pmatrix} a & b & c & d \\ \alpha & \beta & \gamma & \xi \\ 0 & 2 & 0 & 1 \\ 0 & 3 & 1 & 0 \end{pmatrix}^{-1}$$

So y, the total biomass production (or growth rate if the oxygen uptake rate and the carbon-dioxide release rate are used as inputs) can be estimated. Fig. 4 gives the evolution of the estimated biomass versus the measured dry cell weight for example one. Mainly in the exponential growth phase there is a good agreement. However, large deviations were found in the initial growth phase and stationary phase. As already pointed out in the case of k_1a, oxygen uptake estimations are rather poor in the initial phases of the fermentations and therefore, so is also the estimation of the biomass. Moreover, De Buyser and Vansteenkiste (1983) performed a sensitivity analysis on the proposed balance and they found an increasing sensitivity of the ultimate estimation of the biomass on the composition of biomass and substrate as growth proceeds. As some nutrients are depleted in the complex medium, the brutoformula of the substrate can change, causing larger deviations at the end of the fermentation. Also Alford (1978) pointed to some problems encountered in applying chemical balances. Indeed, the substrate is not always one component, especially in the case of a complex medium. Also the substrate-composition is not always known and finally the elementary composition of the cell is submitted to changes during a fermentation. One of the first publications on the use of chemical balances for representing growth comes from Battley (1960). However, he applied this technique only for studying thermodynamical reaction systems. Cooney et al. (1977) builded a growth-estimator based on the balance theory for baker's yeast fermentation. They used two balances, one for representing growth on glucose and one for the conversion of glucose to ethanol. Their method seemed very suited to on-line computer monitoring. Also others have applied this technique to other fermentations with fairly promising results : Goma (1975) in the case of fermentations on hydrocarbons ; Hampel et al. (1981) estimated the substrate-utilization and growth in Arthrobacter sp. fermentations ; Luyben et al. (1981) could monitor different growth phases in Gluconobacter oxydans fermentations in this way ; Madron (1979) was able to implement the technique via matrix-techniques for single cell protein-production on ethanol ; finally, Schwartz and Cooney (1979) monitored the growth of Hansenula polymorpha DL-1 in a continuous culture with methanol as the sole carbon- and energy source. A very important remark to the mentioned examples is that most of them do regard products in their chemical balance. However in this case a supplementary measurement is necessary (four equations, three measurements, four unknowns) as for example substrate utilization is monitored or in some cases a supplementary balance (usually the enthalpy balance) is used. In this respect, it is worthwhile to point to some recent developments. It seems that determining just sufficient variables as are necessary for the estimation of the unknown results in a less stable calculation where as measuring at least one more variable not only delivers more accurate estimations but also make a determination of the errors on the measurements possible. In this way consistency of the data can be checked. A whole theory around this "overdetermined system" technique has been developed. Initially the theory has been described for chemical processes (Van der Grinten and Lenoir, 1973 ; Murthy, 1973, 1974 ; Madron and Vanecek, 1977). Madron and Vanecek introduced in 1977 the principles in fermentation technology. Madron (1979) and De Kok and Roels (1980) showed some good examples as e.g. the correction of RQ-estimations, a parameter very important in the process-control of baker's yeast fermentation. Another group of researchers proposed a similar application of mass- and energy balances in order to detect regularities and irregularities in analyses of fermentations (Erickson et al., 1978a, 1978b, 1979 ; Erickson, 1980 ; Ferrer and Erickson, 1979 ; Minkevich, 1983 ; Solomon and Erickson, 1981 ; Solomon et al., 1982 ; Sobotka et al., 1983). Hereby they introduced the concept of the available electron balance, which seemed in some cases, depending on the nitrogen source, to be equal to an energy balance and which may prove to be valid. Finally, Wang and Stephanopoulos (1983) proposed a multi-dimensional chi-square test for detecting the

presence of gross errors and also a strategy to establish the source
of these errors which previous methods could not determine unambi-
guously. Finally, the usefulness of these "overdetermined system"
techniques may become especially significant as more of the previously described "hardware" sensors enable us to take more measurements.
Therefore, the development of "hardware" sensors does not necessa-
rily mean the redundancy of "net-effect" sensors !

Problems in optimization and control studies

Just as in the chemical industry, control of the process itself
in order to obtain a better return is one of the prime objectives.
The goal is to look for best time profiles rather than constant values
for the control inputs such as temperature, pH, inlet flow rate and
outlet flow rate. Unfortunately nowadays, the problem is far from
resolved in the context of biotechnological processes. The state
of the art is discussed in the following paragraphs.

The easiest form of optimization is steady state optimization,
mainly applied in the case of continuous bioreactors where steady-
state conditions hold. Because optimal steady-state conditions are
looked after the setpoints are constant. This simplifies the problem.
A first approach is to start from a mathematical model and to rely
on special computer-aided calculation schemes. Often dynamic pro--
gramming is applied. The success of the approach is highly tied to
the validity of the mathematical model. The mathematical models in
the fermentation field are at least to a certain degree able to
describe the stationary conditions of a growing micro-organism.
If the product formation path is biochemically known, the mathematical optimization procedure will have a solution that is at least
an improvement compared with a traditional approach. If no model
is available, one can still try to implement "on-line optimization"
through experiment. In this method, control variables are continuously changed and the process-outcome monitored. If this change was
advantageous, one continues in this direction, untill an optimum
is reached. Basically one utilizes the same optimization scheme,
the only difference resides in the fact that in the former case
one relies on the model in order to compute the cost of certain
specific control inputs while in the latter case the system itself
is used in an experiment to compute that cost. In the second approach
one avoids the problem of model validity. However, it is extremely
time consuming due to the slow adaptation of fermentation processes
to new conditions. In addition one has to face the problem of con-
tamination and eventual strain mutation during this long process.

Dynamic optimization is much more involved. It applies for batch
but especially for fed-batch fermentation. Here the experimental
approach is not feasible, so one always needs a mathematical representation of the process. As optimal time control profiles has
to be found one has to apply variational calculus or the celebrated
principle of Pontryagin (Pontryagin et al., 1962).

The procedure has yet several times been used in practice. Foulard and Bourdand (1975) report the optimization of erythromycine production by looking for optimal temperature and pH-profiles.
Constantinides (1970) has obtained in this way optimal temperature profiles for penicilline production. D'Ans et al. (1972) report optimization of bacterial growth where inlet flow rate during continuous cultivation was the control parameter to be found. Sometimes the procedure can be implemented by a feedback loop.
This solution has to be advocated for reasons of stability. Unfortunately dynamic optimization has not been overly successful at the present. The major reason is the poor validity of the models involved. The common model for microbial growth is the Monod-law and substrate consumption is accounted for by a linear equation (model of Pirt). Product formation is described by the Luedeking and Piret equation. One obtains :

$$\frac{dx}{dt} = \frac{\mu_m S x}{K_S + x} \qquad\qquad x(t_o) = x_o$$

$$\frac{dS}{dt} = - \frac{1}{Y_{x/S}} \cdot \frac{dx}{dt} + m_S x \qquad\qquad S(t_o) = S_o$$

$$\frac{dP}{dt} = \frac{1}{Y_{x/P}} \cdot \frac{dx}{dt} + m_P x \qquad\qquad P(t_o) = 0$$

x   : biomass (gr. DCW/1.)

S   : substrate concentration (gr. substrate/1.)

μ_m : maximum specific growth rate (1/h)

K_S : saturation constant (gr. substrate/1.)

$Y_{x/S}$: yield coefficient for substrate (gr. DCW/gr. substrate)

m_S : maintenance coefficient (gr. substrate/gr. DCW.h)

$Y_{x/P}$: yield factor for product (gr. product/gr. DCW)

m_P : production factor (gr. product/gr. DCW.h)

The constants μ_m, K_S, $Y_{x/S}$, m_S, $Y_{x/P}$, m_P are functions of the control variables in a manner that is badly known. Usually one defines those constants as polynomials in the control variables and one tries to obtain the parameters, that appear, through independent measurements. A suitable cost function is defined and an optimization procedure is applied.

Achievements has been rather disappointing. In the first place there is some discussion about the correctness of this model (Spriet, 1982). Furthermore, the model is too simple to apply for industrial contexts where the substrates are often badly known. Finally the equation is mainly suitable for steady-state growth or balanced growth conditions and not for transients which are of main interest for dynamic optimization.

Lately, there have been some important contributions to model building in order to improve the mathematical representation of the microbial process. First structured models have been proposed (Roels, 1983). The biomass is thought to be made up of different compartments and the substrate is also modelled in more detail. Two problems arise here. First validation is difficult because of the lack of sensors. It is indeed difficult to set-up proper experiments in order to fully explore the potentials of new more complicated mathematical models. Second it is not known yet how much detail one has to incorporate in the model so that transients can be properly accounted for. New findings concerning the thermodynamic efficiency of microbial growth and product formation may well shed a new light on this problem. Thirdly it is an additional question to find how the control variables are mathematically related to the parameters of those structured models.

A second approach, in the quest for more faithful and adequate control profiles has been the extension of deterministic models to stochastic ones. The uncertainty in the model is explicitly accounted for through the addition of a Gaussian white noise term. For the growth rate equation e.g. the following stochastic differential equation has been proposed :

$$\frac{dx}{dt} = \frac{\mu_m S}{K_S + S} x + w(t) \qquad\qquad x(t_o) = x_o$$

$w(t)$: Gaussian white noise with mean 0 and variance V_x

The model has been analyzed in order to see whether the heuristic modification did not result in model properties in contradiction with the basic features of any growth process. Fortunately the behaviour of the probability density of x as a function in time is sound and acceptable. In a second step stochastic optimal control has to be applied. It is not known yet whether the stochastic approach will yield more insight or better control profiles. What can be said however is that the manner of obtaining stochastic models can be improved considerably. There are basically two approaches. First, one can start from a corpuscular viewpoint of the fermentation process. Such viewpoint is inherently stochastic. The problem consists in reducing the complexity in a physically acceptable way in order to obtain a model of workable dimensions. The second approach would consist in a critical investigation into the model assumptions that underly the macroscopic deterministic models. Those model parts that are derived from save principles could be retained as deterministic - e.g. mass balances are certain to hold - other model details that have been established on more hypothetical grounds could be made stochastic - e.g. rate equations or parameters that are subject to external disturbances. The approach mentioned here is not that involved. Unfortunately one quickly runs into stochastic models that have too complex a structure. The stochastic control field has been used to models of the following type :

$$\frac{dx}{dt} = f(x,u,t) + g(x,u,t)\ w(t)$$

Unfortunately the slightest extensions in the biological field yield equations that are not additive as the equation mentioned. So either the mathematical control field is extended or we have to simplify a stochastic biological equation in an acceptable manner until the structure is sufficiently simple to handle in a control context.

A third solution to control is the so-called adaptive control solution. In principle one avoids the modelling problem by accepting a time variable linear model which parameters are estimated on-line. In a sense a parametric model is adapted continuously to the plant dynamics hoping that the adaptation will be adequate for control actions derived from the "black box" model. Bastin et al. (1982a, 1982b) ; Meiners and Rapmundt (1981) have considered the problem. Adaptive control has been very effective yet to a large number of engineering problems. Unfortunately, the problem is not yet completely resolved in the fermentation context. Indeed recent findings by Bastin using simulation on classic fermentation models has shown that in a variety of cases the adaptive scheme fails. The biological system is very non-linear and the control loop turns out to be unstable. All hope has not to be lost. Indeed there is the possibility to extent the adaptive procedure to black-box models that are non-linear in the equation but linear in the parameters. Such models seem to be more suitable in the fermentation context and in some instance one has been able to ensure stability. Remains to be seen how those new approaches will actually perform on a true fermentation process.

SOFTWARE

This section discusses the implementation of the software when computers are interfaced to fermentors. We shall not discuss the special software-techniques applied to modelling, simulation and optimization studies. All these can be found in the book of Spriet and Vansteenkiste (1982).

Generalities

When acquiring a computer, a number of standard programs are always provided by the manufacturer. These programs are meant to facilitate the use of the machine and the development of new programs by the customer. One of the most important of these programs is the Executive. It is the program who decides what and where to run. Different Executives can be obtained depending on the purpose of the computer. In the fermentation context, one can easily draw following conclusions :

- The microbial world is very rich and a wide variety of processes occur. Programs must be able to cope with a wide range of circumstances and thus modularity in program design is required so that a package can be assembled for each separate case.
- Data acquisition and process control require on-line and real-time operations, the Executive must be able to take time directly into account.
- As different tasks need to be done at different levels of time priority, it is useful if different tasks may compete for computer resources on the basis of their priority.

Useful Executives should thus provide the opportunity of "multi-programming" and "real-time handling". This means that different tasks can be at work simultaneously but that the specific task that has the highest time priority acquires the processor for immediate execution. After termination, the other running tasks can be checked and the one with the highest priority restarted.

It is in practice not difficult to find a good executive, the real problem however is to develop the necessary software that is directly related to the fermentation process.

Computer activities can be divided into on-line and off-line process analysis tasks, on-line and off-line proces control tasks and operator-machine interfacing.

On-line and off-line process analysis

Figure 6 represents the whole software package for on-line process analysis. One of the simplest functions which can be fulfilled by a computer interfaced to a process is data logging. Values indicated by different sensors are read periodically. The digital quantities, coming from the ADC, are converted by programs into practical units corresponding to the physical quantities. This is done basically through data-acquisition-software of which the major part is a "driver" which handles the computer output and input ports. It is worthwhile to mention that some preliminary tests are required to gain insight into the nature of the noise processes on the fermentor to computer links. If necessary several statistical filters exist for processing noisy data and even simple data averaging may provide adequate data processing. As a result of the data logging process, values of the various process variables can be print out at regular intervals, e.g. every 15 minutes. In this way a record is obtained, which can be consulted by a human operator to follow the process.

Typically for fermentation process software is the existence of a somewhat higher software level in the total hierarchy that is completely dedicated to the net-effect sensor. Several variables are properly interrelated in order to provide a clue variable into the dynamics of the process itself. At this point all instantaneous estimations of the important process variables are now available.

222

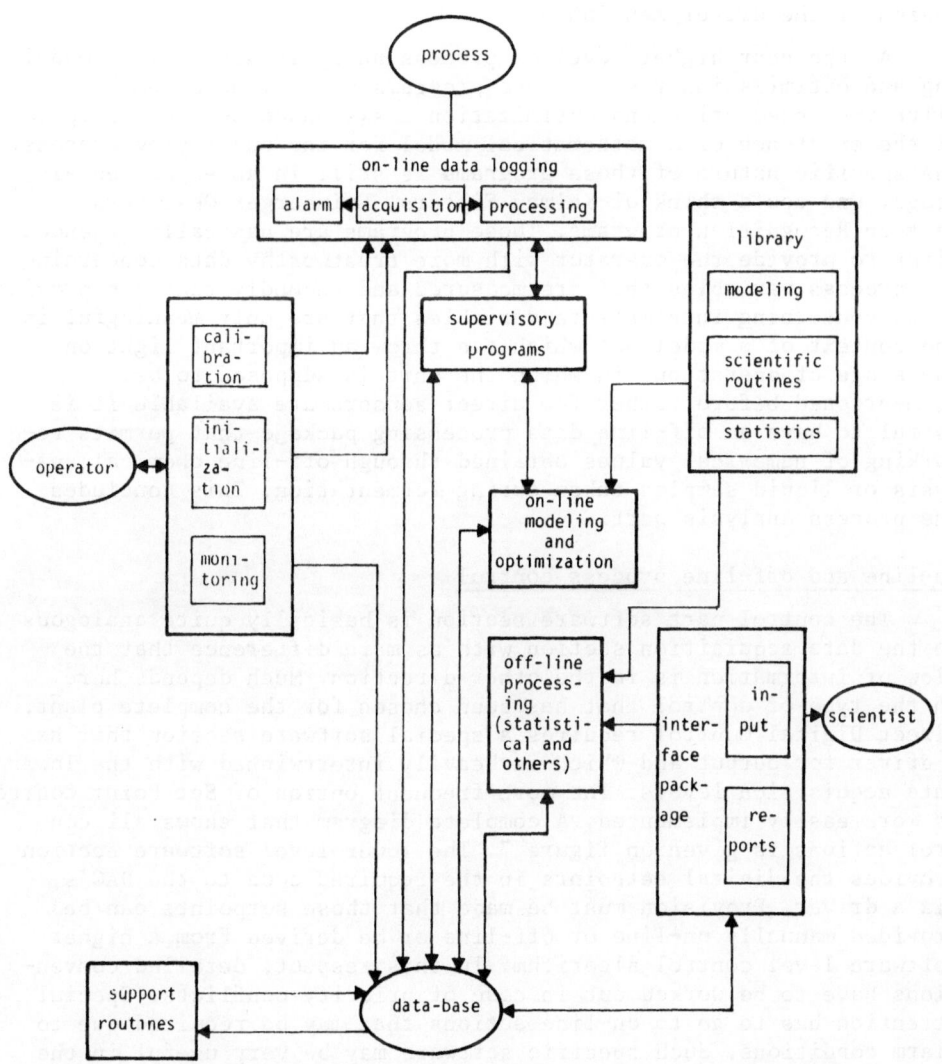

Figure 6 : Program-package for process-control and off-line processing

223

At this point several provisions for detecting alarm-conditions may be included. An alarm condition indicates that something has gone wrong : fermentor malfunction, sensor malfunction, electrical connection malfunctions. It is wise to buffer the alarm handling routines for occasional noisy spikes. This may be an important issue in the design of the driver mentioned.

At the next higher level of process analysis are on-line modelling and optimization tasks. Those programs perform more complicated filtering, prediction and optimization tasks based on the assumption of the existence of a mathematical model for the underlying process. The specific nature of those programs is still in an experimental stage. One could think of Kalman Filters, Luinberger Observers, Pattern Recognition programs. Those programs are basically intended first to provide the operator with more trustworthy data concerning the process variables that are measured and secondly they can provide clues concerning intermediate variables that are only meaningful in the context of a model but which can throw an important light on the stage of operation, in which the unit is supposed to be. As mentioned before rather few direct sensors are available it is useful to have an off-line data processing package that permits reworking of numerical values obtained through off-line chemical analysis of liquid samples taken during fermentation. This concludes the process analysis part.

On-line and off-line process control

The control part software section is basically quite analogous to the data acquisition section with as main difference that the flow of information is in the other direction. Much depends here on the type of control that has been chosen for the complete plant. Direct Digital Control requires a special software section that has a driver for output and which is heavily intertwinned with the lower data acquisition levels. The more frequent option of Set Point Control is more easily implemented. A complete diagram that shows all control actions is given on figure 7. The lower level software section provides the digital setpoints in the required code to the DAC's via a driver. Provision must be made that those setpoints can be provided manually on-line or off-line or be derived from a higher software level control algorithm. In this respect, detailed conventions have to be worked out in case of priority conflicts. Special attention has to go to on-line actions that may be required due to alarm conditions. Such specific software may be very useful in the complete automation of the plant where malfunctioning of a unit is properly circumvented by a well-prepared computer-action.

Operator-machine interfaces

An interface console is a major and important part of the whole apparatus and software has to be provided in order to make the man-machine interaction as user-friendly as possible.

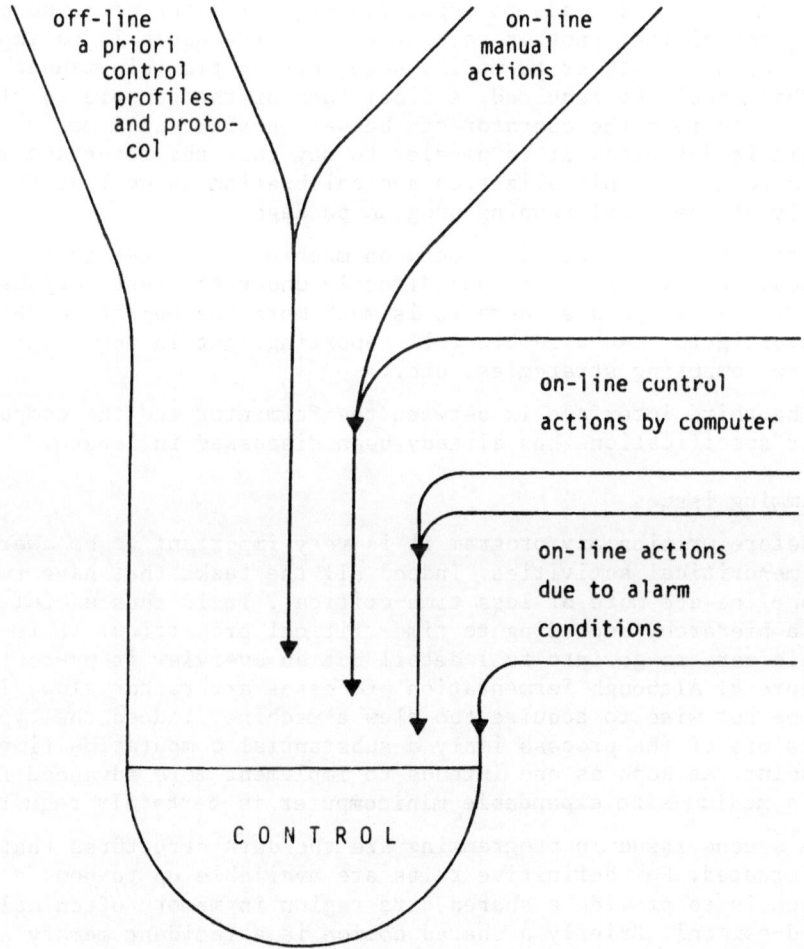

off-line
a priori
control
profiles
and proto-
col

on-line
manual
actions

on-line control
actions by computer

on-line actions
due to alarm
conditions

CONTROL

Figure 7 : Information-flow diagram for control

The most important interface deals with information exchange between the process operator and the process via the digital computer. In its basic mode there must be the possibility to display instantaneous values of the distinct variables in a diagram form, or to display the process in a snapshot picture, or to show relationships between variables or to provide graphical output of the time course of several variables between two specified moments. Provisions should be made to input raw data from the off-line processing activities. Another task of the input console is to input records of setpoints or to modify setpoints in flexible manner whenever this should be required. A final task of the console is alarm-reporting so that the operator can be set on alert if a malfunctioning unit is detected. It is needles to say that the interface must be able to assist initialization and calibration as well as the assembly of the total running program package.

Often a second interface between machine and investigator is provided. This interface is not directly under the restrictions of time-critical responses. Here it is much more the objective that the investigator can assemble full reporting, put in new programs, test new computing strategies, etc..

The third interface is between the fermentor and the computer and its specifications has already been discussed in length.

Programming issues

Before writing any program it is very important to be aware of the time-critical activities. Indeed all the tasks that have to be done on-line are more or less time-critical. It is thus useful to build a hierarchy according to time-critical properties. It is not possible here to go into full detail but an overview is presented on figure 8. Although fermentation processes are rather slow, it would be not wise to acquire too slow a machine. Indeed the typical constraints of the process imply a substantial computation time per data-point. As soon as one intends to implement more advanced features a medium-size expandable minicomputer is certainly required.

A second issue in programming are the data structures that have to be created. Few definitive rules are available up to now. A safe approach is to provide a shared data region in memory often called 'shared-common'. Briefly a shared common is a resident memory area which is accessible by different tasks. It is mainly used to transfer data in an orderly manner from one program to another without using mass storage devices. This is particularly important in time critical activities. Besides a shared common, there should be a well organized data base. This data base gathers on-line and off-line data in an orderly manner and provides a valuable data pool for all kind of further operations.

A third issue in programming is modularity. At the present there are no well-established rules, that provide a given structure

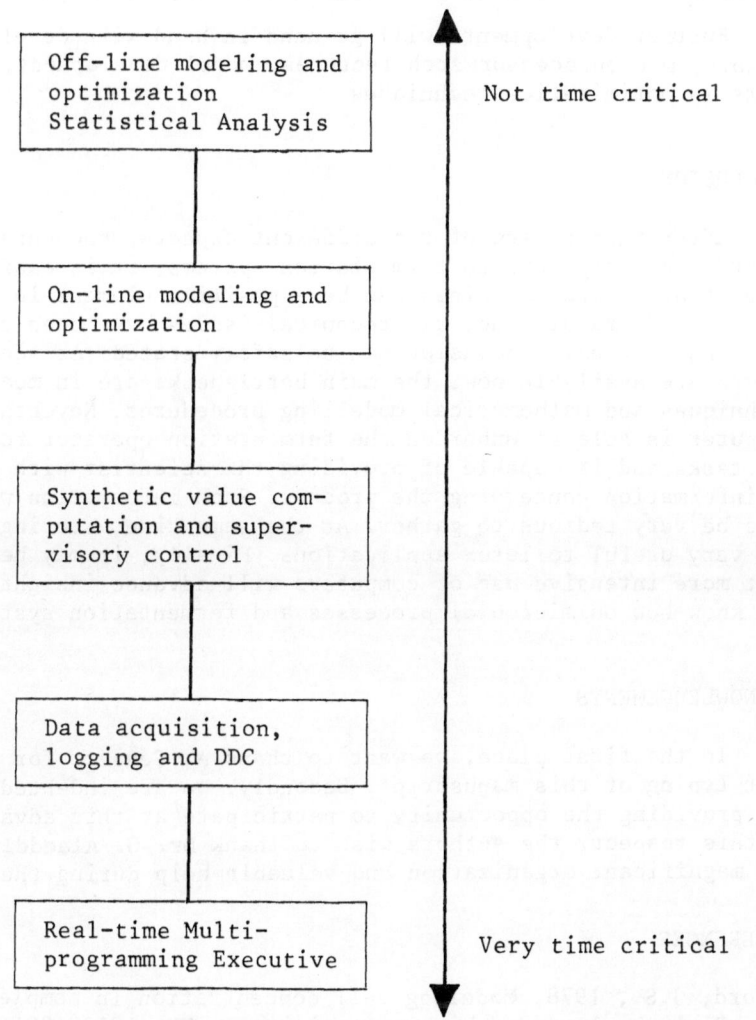

Fig. 8 : Heirarchical order according to time critical requirements

for modularity that is directly linked and properly related to fermentation software needs.

Practical implementations are available, still in a badly documented form. To be mentioned are the work of Rolf and Lim (1982), Rolf et al. (1982), Meiners and Rapmundt (1983), and our personal software as described in De Buyser and De Wael (1981).

Further developments will go hand in hand with developments in sensors, microprocessor architectures, model development, and improvements in optimization techniques.

CONCLUSION

After this review of the different aspects, encountered in interfacing computers to fermentation systems, it is easily understood that digital machines can be a positive element in fermentation industry. Certainly, not all technical issues have been considered here, but a final conclusion can be safely stated. As adequate computers are available now, the main bottlenecks are in measurement techniques and mathematical modelling procedures. Nevertheless, the computer is able to unburden the fermentation operator for most of his tasks and is capable of providing the scientist with a wealth of information concerning the process. Such information would otherwise be very tedious to gather. At the same time, storing facilities are very useful to later applications. Finally, it may be expected that more intensive use of computers will advance insight and improve the know-how on microbial processes and fermentation systems.

ACKNOWLEDGEMENTS

In the first place, we want to thank A. Gevaert for the excellent typing of this manuscript. Secondly, we are indebted to NATO for providing the opportunity to participate at this advanced course. In this respect, the authors wish to thank Dr. G. Alaeddinoğlu for the magnificant organization and valuable help during the course.

REFERENCES

Alford, J.S., 1978, Modeling cell concentration in complex media, Biotechnology and bioengineering,vol. XX, 1873-1881.
Bastin, G., Dochain, D., Haest, M., Installé, M. and Opdenacker, P., 1982a, Modeling and adaptive control of a continuous anaerobic

fermentation process, IFAC-workshop on Modelling and Control of biotechnical processes, Helsinki, Finland, august 17-19.

Bastin, G., Dochain, D., Haest, M., Installé, M. and Opdenacker, P., 1982b, Identification and adaptive control of a biomethanization process, IFIP Working Conference on modelling and data analysis in biotechnology and medical engineering, University of Ghent, August 31 - September 2.

Battley, E.H., 1960, Growth-Reaction Equations for Saccharomyces cerevisiae, Physiologia Plantarum, vol. 13, 192-203.

Bayer, K. and Fuehrer, F., 1982, Computer coupled calorimetry in fermentation, Process biochemistry, July/august, 42-45.

Bergveld, D. and De Rooij, N., 1979, From conventional membrane electrodes to ionsensitive field-effect transistors, Med. & Biol. Eng. & Comput., 17, 647-654.

Bernard, A., Cordonnier, M. and Lebeault, J.M., 1983, A DDC pilot plant system for fermentation control, Process Biochemistry, may/june.

Beyeler, W., Einsele, A. and Fiechter, A., 1981, Fluorometric studies in bioreactors : method and applications, Second European Congress on Biotechnology, 5-10 april, Eastbourne.

Blachère, H. and Jamart, G., 1969, A flow cell photometer for bacterial growth monitoring, Biotechnology and bioengineering, vol. XI, 1005-1010.

Constantinides, A., Spencer, J. and Gaden, E., 1970, Biotechnology and Bioengineering, vol.12.

Cooney, C.L., Wang, D.I.C. and Mateles, R.I., 1968, Measurement of heat evolution and correlation with oxygen consumption during microbial growth, Biotechnology and Bioengineering, vol. XI, 269-281.

Cooney, C.L., Wang, H.Y. and Wang, D.I.C., 1977, Computer-aided fermentation monitoring and diagnostics, US/USSR Sem. on Measurements in Fermentation Processes, Philadelphia, U.S.A., august 12-14.

Cooney, C.L., Wang, H.Y. and Wang, D.I.C., 1977, Computer-aided material balancing for prediction of fermentation parameters, Biotechnology and bioengineering, vol. XIX, 55-67.

Croset, M., 1982, Chemically sensitive ionic devices, Summercourse on solid-state sensors (W. Sansen and J. van der Spiegel), K.U.L., Belgium.

Dahod, S.K., 1982, Redox potential as a better substitute for dissolved oxygen in fermentation process control, Biotechnology and bioengineering, vol. XXIV, 2123-2125.

Danielsson, B., Mandenius, C.F., Winquist, F., Mattiasson, B. and Mosbach, K., 1981, Enzyme thermistor applications in biotechnology, 2nd European Congress of Biotechn., p.119, 5-10 April.

Danielsson, B., Winquist, F., Mosbach, K. and Lundström, I., 1983,
    Enzyme transistors, Biotech 83, On-line Publications Ltd.,
    679-688.
D'Ans, G., Gottlieb, D. and Kokotovic, P., 1972, Optimal Control of
    Bacterial Growth, Automatica 8, Pergamon Press.
De Buyser, D.R. and De Wael, L.A., 1981, On-line software for process
    monitoring, Internal report, University of Ghent, Belgium.
De Buyser, D.R. and Vansteenkiste, G.C., 1983, Building an on-line
    sensor for biomass through modelling, First European Simulation
    Congress ESC 83 (Ed. W. Ameling), 390-396.
De Kok, H.E. and Roels, J.A., 1980, Method for the statistical treat-
    ment of elemental and energy balances with application to steady-
    state continuous culture growth of Saccharomyces cerevisiae
    CBS 426 in the respiratory region, Biotechnology and bioengineer-
    ing, vol. XXII, 1097-1104.
Enfors, S.O., 1981a, An enzyme electrode for control of glucose con-
    centration in fermentation broths, Second European Congress of
    Biotechnology, 141, 5-10 april, Eastbourne.
Enfors, S.O., 1981b, Oxygen-stabilized enzyme electrode for D-glucose
    analysis in fermentation broths, Enzyme Microb. Technol., vol.3,
    29-32.
Enfors, S.O. and Dostalek, M., 1975, Monitoring of dosing of liquids
    in laboratory scale fermentation, Process Biochemistry,
    july/august.
Erickson, L.E., 1980, Analysis of microbial growth and product forma-
    tion with nitrate as nitrogen source, Biotechnology and bioen-
    gineering, vol. XXII, 1929-1944.
Erickson, L.E., Minkevich, I.G. and Eroshin, V.K., 1978a, Application
    of man and energy balance regularities in fermentation, Bio-
    technology and bioengineering, vol. XX, 1595-1621.
Erickson, L.E. and Viesturs, U.E., 1978b, Application of man and
    energy balance regularities to product formation, Biotechnology
    and bioengineering, vol. XX, 1623-1638.
Erickson, L.E., Minkevich, I.G. and Eroshin, V.K., 1979, Utilization
    of mass-energy balance regularities in the analysis of continu-
    ous culture data, Biotechnology and bioengineering, vol. XXI,
    575-591.
Ericksson, R.K., 1981, Observations of microbial substrate utiliza-
    tion via the heat produced, Second European Congress of biotech-
    nology, 138, 5-10 april, Eastbourne.
Esener, A.A., Veerman, T., Roels, J.A. and Kossen, N.W.F., 1982,
    Modeling of bacterial growth ; formulation and evaluation of a
    structured model, Biotechnology and bioengineering, vol. XXIV,
    1749-1764.
Fazel-Madjlessi, J. and Bailey, J.E., 1979, Analysis of fermentation
    processes using flow microfluorometry : single-parameter obser-
    vations of batch bacterial growth, Biotechnology and bioengineer-
    ing, vol. XXI, 1995-2010.
Ferrer, A. and Erickson, L.E., 1979, Evaluation of data consistency
    and estimation of yield parameters in hydrocarbon fermentations,
    Biotechnology and bioengineering, vol. XXI, 2203-2233.

Fish, N.M., Vardar, F. and Lilly, M.D., 1981, Effect of dissolved gas concentrations on microbial product formation, Second European Congress of biotechn., 77, 5-10 april, Eastbourne.

Foulard, C. and Bourdand, D., 1975, Optimisation de procédes de fermentation discontinus, in "Automatisation des processus de fermentation", AFCET.

Fraser, G., 1983, Computer control of fermentation processes, Biotech 83, Online Publications Ltd., 283-293.

Goma, G., 1975, Contribution à l'étude des fermentations sur hydrocarbones, Transfert de matière - lois de croissance, Thèse présentée à l'université Paul Sabatier de Toulouse pour obtenir le grade de docteur d'état, 1-231.

Goma, G. and Ribot, D., 1975, Procédé de mesure des chaleurs de réaction applicable à la détermination des paramètres de la croissance microbienne et des productivités des réacteurs biologiques, Brevet (France) 75-14927.

Guilbault, G.G., 1980, Enzyme electrode probes, Enzyme Microb. Technol., vol.2, 258-264.

Hampel, W., 1980, Bildung von α-Mannosidase durch Arthrobacter Monatshefte für Chemie, 111, 443-457.

Hampel, W., Wöhrer, W., Bach, H.P. and Röhr, M., 1979, Computer unterstützte analyse und steuerung von fermentationen, Mitteilungen der versuchsstation für das gärungsgewerbe in Wien, n°1/2.

Hampel, W., John, E. and Roehr, M., 1981, Minicomputer application in fermentation research : analysis and control of enzyme formation in Arthrobacter sp., Second European Congress of Biotechnology, 5-10 april, Eastbourne, England, Abstract of communications, p.147 and personal letter.

Heijnen, J.J., 1981, Application of the macroscopic electric charge balance in fermentation modeling, Second European Congress of Biotechnology, 5-10 april, Eastbourne, p.92.

Heinzle, E., 1981, Continuous on-line measurement of process variables in biological reactors, XXII cycle de perfectionnement en genie chimique, thème : "le genie biochimique", 25-27 nov. Brussel, Belgium.

Heinzle, E., Dunn, I.J. and Bourne, J.R., 1981, Continuous measurement of gases, dissolved gases and volatiles during fermentations using mass spectrometry, Second European Congress of biotechnology, p.24, 5-10 april, Eastbourne.

Higham, E.H., 1983, Practical issues for microprocessors in process measurements, Symposium 'Impakt van de mikro-elektronica op de primaire meetelementen', 6 oktober 1983.

Hill, F.F. and Thommel, J., 1982, Continuous measurement of the ammonium concentration during the propagation of baker's yeast, Process Biochemistry, September/October, 16-18.

Holmberg, A., 1981, A systems engineering approach to biotechnical processes - experiences of modelling, estimation and control methods, Acta Polytechnica Scandinavia, mathematics and computer science series n°33, 1-46.

Holmberg, A., Sievänen, R. and Carlberg, G., 1980, Fermentation of
    Bacillus thuringiensis for exotoxin production : process analysis
    study, _Biotechnology and bioengineering_, vol.XXII, 1707-1724.

Humphrey, A.E., 1977, The use of computers in fermentation systems,
    Process Biochemistry, march, 19-25.

Jarai, M., 1972, Oxygen transfer in the fermentations of primary and
    secondary metabolites, Proc. IV IFS : Ferment. Technol. Today,
    97-103.

Karube, I. and Suzuki, S., 1983, Biosensor for fermentation and en-
    vironmental control, Biotech 83, Online Publications, 625-632.

Kell, D.B., 1980, The role of ion-selective electrodes in improving
    fermentation yields, _Process Biochemistry_, january, 18-23.

Kempe, E. and Schallenberger, W., 1983, Measuring and control of
    fermentation processes : part 1, _Process Biochemistry_, december,
    7-12.

Kernevez, J.P., Konate, L. and Romette, J.L., 1983, Determination of
    substrate concentrations by a computerized enzyme electrode,
    Biotechnology and bioengineering, vol. XXV, 845-855.

Kjaergaard, L. and Joergensen, B.B., 1979, Redox potential as a state
    variable in fermentation systems, Biotechnology and bioengineer-
    ing symposium, n°9, 85-94.

Le Duy, A. and Samson, R., 1982, Testing of an ammonia ion selective
    electrode for ammonia nitrogen measurement in the methanogenic
    sludge, _Biotechnology letters_, vol.4, n°5, 303-306.

Lee, C. and Lim, H., 1980, New device for continuously monitoring the
    optical density of concentrated microbial cultures, _Biotechnology
    and bioengineering_, vol. XXII, 639-642.

Leisola, M., Ojamo, H. and Kauppinen, V., 1979a, Automatic monitoring
    of protease activity during fermentation, _Enzyme Microb. Technol._,
    vol.1, 51-52.

Leisola, M., Virkkunen, J., Karvonen, E. and Meskanen, A., 1979b,
    Automatic cellulase assay in computer coupled pilot fermentation,
    Enzyme Microb. Technol., vol.1, 117-121.

Leisola, M., Ojamo, H., Kauppinen, V., Linko, M. and Virkkunen, J.,
    1980, Measurement of α-amylase and gluco-amylase activities pro-
    duced during fermentation, _Enzyme Microb. Technol._, vol.2,
    121-125.

Liu, C.C. and Chen, A.K., 1982, Potentiometric quantitation of bio-
    logical substrates using gel-immobilized oxidoreductases,
    Process Biochemistry, september/october, 12-14.

Lock, M.A and Ford, T.E., 1983, Inexpensive flow microcalorimeter
    for measuring heat production of attached and sedimentary aqua-
    tic micro-organisms, _Applied and environmental microbiology_,
    vol.46, n°2, 463-467.

Lowe, C., Goldfinch, M., Lias, R., 1983, Some novel biomedical
    biosensors, Biotech, Online Publications Ltd., 633-641.

Luyben, K.Ch.A.M., Tramper, J. and Olieman, J.J., 1981, Monitoring
    different growth phases in _Gluconobacter oxydans_ fermentation,
    Sec. European Congress of Biotech., 5-10 april, Eastbourne,
    139.

232

Madron, F., 1979, Material-balance calculations of fermentation processes, Biotechnology and bioengineering, vol. XXI, 1487-1490.

Madron, F. and Vanecek, V., 1977, Statistical adjustment of material balance of a chemical reactor, Collection Czechoslov. Chem. Commun., 42.

Mandenius, C.F., Danielsson, B. and Mattiasson, B., 1981, Control of the substrate concentration in an ethanol fermentation by an enzyme thermistor, Second European Congress of Biotechnology, p.143, 5-10 april, Eastbourne.

Marison, I. and von Stockar, U., 1983, The use of a new heat flux calorimeter for the measurement of heat evolved during microbial growth, Biotech 83, Online Publications, 947-959.

Meiners, M. and Rapmundt, W., 1981, Adaptive and/or optimal control in fermentation processes ?, Second European Congress of Biotechnology, p.91, 5-10 april, Eastbourne.

Meiners, M. and Rapmundt, W., 1983, Some practical aspects of computer applications in a fermentor hall, Biotechnology and bioengineering, vol. XXV, 809-844.

Minkevich, I.G., 1983, Mass-energy balance for microbial product synthesis - biochemical and cultural aspects, Biotechnology and bioengineering, vol. XXV, 1267-1293.

Mosbach, K., Mandenius, C.F. and Danielsson, B., 1983, New biosensor devices, Biotech 83, Online publications Ltd., 665-678.

Mou, D.G. and Cooney, C.L., 1976, Application of dynamic calorimetry for monitoring fermentation processes, Biotechnology and bioengineering, vol. XVIII, 1371-1392.

Murthy, A.K.S., 1973, A least-squares solution to mass balance around a chemical reactor, Ind. Eng. Chem. Process Des. Develop., vol.12, n°3.

Murthy, A.K.S., 1974, Material balance around a chemical reactor,II, Ind. Eng. Chem. Process Des. Develop., vol.13, n°4.

Nagai, S., 1979, Mass and energy balances for microbial growth kinetics, Advances in biochemical engineering, n°11, 49-83.

Nyeste, L., Szigeti, L., Veres, A., Pungor, E., Kurucz, I. and Hollo, J., 1981, Automated Fermentation equipment, II. Computer-fermentor system, Biotechnology and bioengineering, vol.XXIII, 405-417.

Nyiri, L., 1972, A philosophy of data acquisition analysis and computer control of fermentation processes, Developments in Industrial Microbiology, 13, 136.

Nyiri, L.K., 1973, Application of computers in biochemical engineering, In 'Advances in Biochemical Engineering', 2, 49-95.

Nyiri, L.K., Toth, G.M. and Charles, M., 1975, On-line measurement of gas-exchange conditions in fermentation processes. Biotechnology and bioengineering, vol.XVII, pp.1663-1678,

Ohashi, M., Watanabe, T., Ishikawa, T., Watanabe, Y., Miwa, K., Shoda, M., Ishikawa, Y., Ando, T., Shibata, T., Kitsunai, T., Kamiyama, N. and Oikawa, Y., 1979, Sensors and instrumentation Steam-Sterilizable Dissolved oxygen sensor and cell mass sensor for on-line fermentation system control, Proc. Second Internat. Conf. on Comp. Appl. in Ferm. Techn., Biotech. and bioeng. Symp. 9, 103-116.

Pirt, S.J., 1975,"Principles of microbe and cell cultivation", Blackwell scientific publications, Oxford.

Pontryagin, L.S., Boltyanskii, V.G. Camkrelidze, R.V. and Mishchenko E.F. 1962, "The mathematical Theory of Optimal Processes", Interscience Publishers Inc., New York.

Pungor, E., Perley, C.R., Cooney, C.L. and Weaver, J.C., 1980, Continuous monitoring of fermentation outlet gas using a computer coupled MS, Biotechnology letters, vol.2, n°9, 409)414.

Ramirez, A., Durand, A. and Blachère, H.T., 1981, Optimal baker's yeast production in extended fed-batch culture by using a computer coupled pilot fermentator, Sec. Euprean Congress of Biotech., p.26, 5-10 april, Eastbourne.

Roels, J.A., 1983, "Energetics and kinetics in biotechnology, Elsevier Biomedical Press, 1-330.

Roels, J.A. and Kossen, N.W.F., 1978, On the modelling of microbial metabolism, Progress in industrial microbiol., 14, 95-203.

Rolf, M.J., Hennigan, P.J., Mohler, R.D., Weigand, W.A. and Lim, H.C., 1982, Development of a direct digital - controlled fermentor using a microminicomputer hierarchical system, Biotechnology and bioengineering, vol.XXIV, 1191-1210.

Rolf, M.J. and Lim, H.C., 1982, Computer control of fermentation processes, Enzyme and Microbial Technology, vol.4, 370-380.

Sansen, W., 1983, Sensors in silicon, Journal A, vol.24, n°3, 116-122.

Sobotka, M., Votruba, J., Havlik, I. and Minkevich, I.G., 1983, The mass-energy balance of anaerobic methane production, Folia microbiol., 28, 195-204.

Solomon, B.O. and Erickson, L.E., 1981, Biomass yields and maintenance requirements for growth on carbohydrates, Process biochemistry, February/March, 44-49.

Solomon, B.O., Erickson, L.E., Hess, J.E. and Yang, S.S., 1982, Maximum likelihood estimation of growth yields, Biotechnology and bioengineering, vol.XXIV, 633-649.

Solsky, R.L., 1983, Ion selective electrodes in biomedical analysis, CRC Crit. Rev. Anal. Chem., 14, 1-52.

Spriet, J.A., 1980, Struktuur-identifikatie in het licht van patroon-herkenning, Ph-D-thesis, University of Ghent, Belgium.

Spriet, J.A., 1982, Modelling the growth of micro-organisms : a critical appraisal, in : "Environmental Systems Analysis and Management" (S. Rinaldi, ed.), North Holland Publ. Co., Amsterdam, The Netherlands.

Spriet, J.A., Botterman, J., De Buyser, D.R., De Visscher, P.L. and Vandamme, E.J., 1982, A computer-aided noninterfering on-line technique for monitoring oxygen-transfer characteristics during fermentation processes, Biotechnology and bioengineering, vol.XXIV, 1605-1621.

Spriet, J.A. and Vansteenkiste, G.C., 1978, A new approach towards measurements and identification for control of fermentation systems, in : "Simulation of Control Systems", (I. Troch, ed.), North-Holland Publ. Co., Amsterdam, 245-248.

Spriet, J.A. and Vansteenkiste, G.C., 1982, "Computer-aided modeling and simulation", International Lecture Notes in Computer Science, Academic Press, London.

Sumitani, T., Shimizu, N. and Odawara, Y., 1983, Automatic control of fermentor using microcomputer, Biotech 83, Online publications Ltd., 295-306.

Swartz, J.R. and Cooney, C.L., 1979, Indirect fermentation measurements as a basis for control, Biotechnology and bioengineering symp., n°9, 95-101.

Tran, N.D., Romette, J.L. and Thomas, D., 1983, An enzyme electrode for specific determination of L-lysine : a real-time control sensor, Biotechnology and bioengineering, vol.XXV, 329-340.

Turner, A.P.F., 1983, Applications of direct electron transfer. Bio-electrochemistry in sensors and fuel cells, Biotech 83, Online Publications, 643-654.

Vandamme, E.J., De Buyser, D.R., De Visscher, P.L., Spriet, J.A. and Demain, A.L., 1981, Dynamics of gramicidin S and GS-synthetases formation in pH- and aeration-controlled Bacillus brevis ATCC 9999 cultures, Second European Congress on Biotechnology, 5-10 april, Eastbourne.

Vandamme, E.J., Leyman, D., De Buyser, D.R., De Visscher, P.L., Spriet, J.A., Vansteenkiste, G.C., Nimi, O., Poirier, A. and Demain, A.L., 1982, Environmental influences on the dynamics of the gramicidin S fermentation, in : "Peptide Antibiotics - Biosynthesis and Functions", (H. Kleinkauf and H. v.Döhren, eds.), Walter de Gruyter & Co., 117-135.

Van der Grinten, P.M.E.M. and Lenoir, J.M.H., 1973, Statistische procesbeheersing , Prisma-Technica, 50 (Uitgeverij Het Spectrum B.V.), 336-343.

Verduyn, C., Van Dijken, J.P. and Scheffers, W.A., 1983, A simple, sensitive and accurate alcohol electrode, Biotechnology and bioengineering, vol.XXV, 1049-1055.

Vorlop, K.D., Becke, J.W. and Klein, J., 1983, On-line measurement of ethanol with a gas-sensor-dip-electrode, Biotechnology letters, vol.5, n°8, 509-514.

Wang, H.Y., Cooney, C.L. and Wang, D.I.C., 1977, Computer-aided baker's yeast fermentations, Biotechnology and bioengineering, vol. XIX, 69-86.

Wang, N.S. and Stephanopoulos, G., 1983, Application of macroscopic balances to the identification of gross measurement errors, Biotechnology and bioengineering, vol.XXV, 2177-2208.

Wei, C.J., Tanner, R.D., Malaney, G.W. and Charles, M., 1983, An on-line indirect measurement technique for monitoring yeast cell biomass in semi-solid gels, Process biochemistry, march/april, 2-5.

Whaite, P., Aborhey, S., Hong, E. and Rogers, P.L., 1978, Microprocessor control of respiratory quotient, Biotechnology and bioengineering, vol.XX, 1459-1463.

Williams, M.H. and Brain, K.R., 1976, A novel method of growth estimation for suspension cultures, Process biochemistry, may, 41-43.

Wingard, L.B., 1983, Prospects for electrochemical devices and processes based on biotechnology, Biotech 83, Online publications Ltd., 613-624.

Winquist, F., Danielsson, B., Lundström, I. and Mosbach, K., 1981, Biochemical applications of H_2 and NH_3 sensitive Pd-MOS structures, Second Europ. Congr. on Biotech., p.120, 5-10 april, Eastbourne.

Zabriskie, D.W., 1979, Use of culture fluorescence for monitoring of fermentation systems, Biotechnology and bioengineering symp. n°9, 117-123.

Zabriskie, D.W., Armiger, W.B. and Humphrey, A.E., 1976, Applications of computers to the indirect measurement of biomass concentration and growth rate by component balancing, GBF-Monograph Series, n°3.

Zabriskie, D.W. and Humphrey, A.E., 1978, Real-time estimation of aerobic batch fermentation biomass concentration by component balancing, AICHE Journal, vol.24, n°1, 138-146.

INDEX